Understanding Innovation

Series Editors
Christoph Meinel
Larry Leifer

For further volumes:
http://www.springer.com/series/8802

Hasso Plattner • Christoph Meinel • Larry Leifer
Editors

Design Thinking Research

Measuring Performance in Context

TS
171.4

D476
2012

Editors
Hasso Plattner
Christoph Meinel
Hasso-Plattner-Institut
Potsdam, Germany

Larry Leifer
Stanford Center for Design Research
Stanford University
Stanford
California
USA

ISBN 978-3-642-31990-7   ISBN 978-3-642-31991-4 (eBook)
DOI 10.1007/978-3-642-31991-4
Springer Heidelberg New York Dordrecht London

Library of Congress Control Number: 2012948626

Printed on acid-free paper

Springer is part of Springer Science+Business Media (www.springer.com)

Preface

The third volume of our series on Design Thinking Research presents the comprehensive collection of research studies carried out by the HPI-Stanford Design Thinking Research Program. This is a joint program of the Hasso-Plattner-Institute of Design at Stanford University in California and the Hasso-Plattner-Institute (HPI) for IT Systems Engineering in Potsdam, Germany.

The concept of Design Thinking refers to the methods and processes for investigating challenges, acquiring information, analyzing knowledge, and positioning solutions in the design and planning fields. As a style of thinking, it is generally considered the ability to combine empathy, creativity, and rationality in analyzing and fitting solutions to context. The overall goal that stands behind all those activities is to generate innovations.

There are several main factors that are crucial in the development process of Design Thinking. One is definitely the role of teamwork and the impact it has for the outcome of the process. Unified commitment, a collaborative climate, and the reception of support and encouragement by the team members are a prerequisite of all kinds of successful collaboration.

Teamwork requires dedication, coordination, and people skills. Teamwork is usually defined as a joint action by a group of people, in which each person subordinates his or her individual interests and opinions to the unity and efficiency of the group. Certainly, this does not mean that the individual is no longer important; however, it does mean that effective and efficient teamwork goes beyond individual accomplishments. The most effective teamwork is produced when all the individuals involved harmonize their contributions and work toward a common goal which is being put into practice in this program.

A further considerable factor when it comes to Design Thinking and its impact is the element of education. The term "education" implies the notions of learning, preparing for successful leadership and personal achievement, as well as opening the mind to new ideas and methods. Education in its best way should open the mind, and the process itself involves creating and solving our own challenges.

Another meaningful component of this program, as well as of this present book, is the facet of innovation. Innovation means – literally translated – improvement or renewal. This is a universal concept that is as old as mankind and every generation adapts it to their needs and requirements accordingly. To have an idea and to make it feasible is called an "invention." Whether an invention actually turns out to be an innovation only becomes clear when it is actually accepted by the users.

In order to make an invention work, it is crucial that the right instruments are chosen. These have a great and important impact on how Design Thinkers perceive the world around them, how they think, feel, and communicate. It was in 1964 when Marshall McLuhan coined the phrase "The medium is the message."

The actual core message behind the phrase that is constantly quoted and often misinterpreted is that we largely miss the structural changes in our affairs that are introduced subtly, or over long periods of time. Whenever we create a new innovation, many of its features and qualities are fairly obvious to us.

We often know what its advantages and disadvantages might be. But it is also often the case that after a long period of time and experience with the new innovation, we look back and realize that there were some effects of which we were entirely unaware at the onset.

Like the message that McLuhan had in mind when he coined that phrase, it is not primarily the content of the innovation, but the change in interpersonal dynamics that the innovation brings with it. It is crucial to not only look beyond the obvious but to seek the nonobvious changes that are enabled or enhanced by the innovation. In other words, an innovation is anything from which a change emerges. It often takes years or even decades before an invention becomes an innovation and before it becomes apparent to everyone.

I am pleased to witness the development of this book series over the years, and it is a rewarding experience for me to see this third volume coming to life.

May it be a fruitful contribution to the ongoing debate on Design Thinking.

Potsdam/Palo Alto   Hasso Plattner
March 2012

Contents

Design Thinking Research

Christoph Meinel and Larry Leifer

1 Design Thinking as Hunting for Big Ideas, AND, Transporting Them Home to the Organization

1.1 Understanding Innovation Is About Knowing How to Measure It

The path of design thinking is filled with various idea-fragments (Baya 1996; Meinel and Leifer 2011; Sonalkar 2012). One of the core challenges faced by design-thinking teams is to navigate through this sea of fragments, to keep all fragments in their sights while constantly testing alternative configurations in pursuit of a concept worth investing in. During each design thinking operation, we are certain to face challenges. One important finding in all design thinking research projects is that deep design thinking is a synthesis challenge more than it is an ideation challenge. The path is constantly being molded and re-shaped by events and findings. Several steps along the way are sure to be different than on any previous search. Way finding is an adventure that enthralls the design thinker and the researchers who observe closely. In time, we face a moment in which a clear path forward unfolds. It is that point in the cycle where synthesis and divergent thinking, analysis and convergent thinking, and the nature of the problem all come together and resolution has been captured. In design thinking processes there is no

C. Meinel (✉)
Hasso-Plattner-Institut (HPI), für Softwaresystemtechnik GmbH, Prof.-Dr.-Helmert-Str. 2-3, Potsdam 14482, Germany
e-mail: meinel@hpi.uni-potsdam.de

L. Leifer
Stanford Center for Design Research, Stanford University, 424 Panama Mall, Stanford, CA 94305-2232, USA
e-mail: leifer@cdr.stanford.edu

H. Plattner et al. (eds.), *Design Thinking Research*, Understanding Innovation, DOI 10.1007/978-3-642-31991-4_1, © Springer-Verlag Berlin Heidelberg 2012

solitary action or procedure that actually defines the process. There are as many different design processes as there are design thinkers.

In this edition of "Understanding Innovation," you will find research reports about many different lines of study into the nature of the syntheses process with a keen eye to measuring team performance in three distinct scenarios.

1. **Co-located control**: The first is in the semi-controlled environment of co-located teams in "d.school" courses at Stanford University and the Hasso Plattner Institute in Potsdam.
2. **Distributed control**: The second context is that of largely uncontrolled distributed teams in design-courses within Stanford's ME310-Global Network.
3. **Business embedded**: The third context is that of corporate teams in the wild, an environment that is largely unobservable in the research sense while tending to be over-controlled for innovation purpose.

The advantage of the first two scenarios is observability, access to the players, and the relative absence of intellectual property protection concerns. Different experimental settings can easily be implemented and tested, a hypothesis can be formulated, discussed, and affirmed or eventually rejected. The advantage of the third context is the opportunity, however difficult to achieve, for validation in real-life business practice.

The recurrent theme in this edition is the difficulty we face in making performance measurements with robust causality in the face of uncontrolled variables. However, for the committed design researcher, difficulty is a call to duty, an opportunity, an incentive to "get creative", and this is definitively true for all our researchers and PhD students in the HPI-Stanford Design Thinking Research Program. Getting creative in this context is to synthesise new and better experiments with new and more sensitive instruments to sort the "wheat from the chaff," as some of our ancestors once quipped. In ongoing projects and their related research questions, all the research teams involved are looking from various angles and multiple points-of-view in order to triangulate findings in a way that maximizes repeatability while retaining generality.

Our researchers value the notion that they, themselves, must be involved in the design process and related technology in order that they might be in the right place, at the right time, and have right point-of-view to capture insights for connected technology development. They recognize the importance of societal factors and the needs of consumers as individuals and members of society. They are mindful of the link between critical technical functionality and critical user experience. In order to look at those issues from all perspectives and in order to give a multifaceted assessment to the research questions we have been delighted in working with great teams of engineers, computer scientists, humanists, educationists and cultural scientists. In the second edition of our series, we introduced the "Innovation Hunter-Gather" metaphor. Hunters-gatherers are looking for specific solutions to specific problems. This model has emerged from our ways of thinking, discussing, and testing to increase the probability of successful innovation from research, development, and marketing activities. This model has proven useful as a means to communicate the core ideas in design thinking. It emphasizes that there are different roles to be played in the activities of innovating. There is a time to hunt and a time to gather. There are times to seek the next big thing and times to deliver the next big thing. The "Hunter-Gather Model" that we designed describes the dynamics of design thinking. The model that we have elaborated in the second book "Design Thinking Research – Studying Co-Creation in Practice" is about enfolding events, awareness, observation, and real time intervention. We were and still are hunting for ideas that sell. And, another important aspect in our ongoing endeavour is to solve problems, perhaps even to remove problems through design thinking, and to create new products and remarkable services. Design thinking has taken root and continues to help shape our pursuit of understanding. We prefer now to shorten the label and focus on the "Hunter Model." Findings reported in this edition have refined our instruments, focused our attention on critical design team activities, and strengthened their correlation with performance. Still to come will be findings in volume 4 where early evidence suggests that the hunting model may be as or more valuable for the task of bringing big ideas and design thinking "back home" to the organization. We promise surprises to come.

Please join us in our continuing hunt for understanding, efficacy, and validation.

2   The HPI-Stanford Design Thinking Research Program

Having started in 2008, the HPI-Stanford Design Thinking Research Program is fully supported and financed by the Hasso Plattner Foundation.

2.1   Program Vision

This research program engages multidisciplinary research teams. Scientifically they investigate the phenomena of innovation in all its holistic dimensions. Researchers are especially encouraged to develop ambitious, long-term exploration projects that integrate technical, economical, as well as psychological points of view using design thinking tools and methods. The HPI-Stanford Design Thinking Research Program is a dedicated academic research effort focused on understanding the scientific foundations of how and for what reasons the innovation methods of design thinking work.

Beyond descriptive understanding, the goal of this program is to develop metrics that allow assessment and prediction of team performance in order to facilitate real-time performance management. Researchers are encouraged to design, develop and evaluate innovative (analogue and digital) tools that support teams in their creative work. One program focus is on exploring the use of design thinking methods in the field of information technology and IT systems engineering. An important feature of this domain is the need for creative collaboration across spatial and temporal boundaries. In the context of disciplinary diversity, the question of how design thinking methods mesh with traditional engineering and management approaches is addressed. Why does the structure of successful design teams differ substantially from traditional corporate structures?

This program involves multidisciplinary research teams from diverging backgrounds such as engineering, design, a broad range of the humanities as well as sociology and education. A prerequisite to being passionate about developing ambitious, long-term, discovery research projects is the need to expand our understanding of design thinking in its technical, business, and human dimensions.

2.2   Program Priorities

A strong cooperation in the offering of both design thinking education programs is a priority. Both of Schools of Design Thinking at Stanford University and the Hasso Plattner Institute in Potsdam focus on fostering collaboration between researchers of Stanford University and the Hasso Plattner Institute.

Multi-year funding favors projects that set new research priorities for this emergent knowledge domain. Projects are selected based on intellectual merit and evidence of open collaboration. The following guiding research questions are of special interest:

- What are people really thinking and doing when they are engaged in creative design innovation?
- How can new frameworks, tools, systems, and methods augment, capture, and reuse successful practices?
- What is the impact on technology, business, and human performance when design thinking is practiced?
- How do the tools, systems, and methods really work to get the innovation you want when you want it? How do they fail?
- What is the impact on technology, business, and human performance when design thinking is practiced?

The overall topic of **Design Thinking as Hunting for Big Ideas, and, Transporting Them Home to the Organization** leads the way through this book, the third volume in the series *Design Thinking Research*, which is part of the *Understanding Innovation* Series by Springer Publishing.

2.3   Part I: Design Thinking Research in the Context of Co-Located Teams

The second chapter, entitled "Assessing d.learning: Capturing the Journey of Becoming a Design Thinker," by authors **Shelley Goldman, Maureen P. Carroll, Zandile Kabayadondo, Leticia Britos Cavagnaro, Adam W. Royalty, Bernard**

Roth, Swee Hong Kwek and Jain Kim explores the relationship of learning design thinking and assessing that progress. It addresses the question: How can one understand what is learned in design thinking classes and how might assessments contribute to that process in authentic ways? The study follows a reciprocal research and design methodology where basic research and the design of assessment solutions are ongoing, reciprocal, and related to each other in organic ways. The research team discovered that the learning of design thinking dispositions and mindsets is an emergent journey – with various levels of sophistication, transformation, application, and integration. They introduce the concept of *mindshifts* to represent the developing and nascent epistemological viewpoints and instincts that are strengthened while becoming a design thinker.

The third chapter by **Birgit Jobst, Eva Köppen, Tilmann Lindberg, Josephine Moritz, Holger Rhinow, Christoph Meinel** bears the title "The Faith Factor in Design Thinking: Creative Confidence Through d.school Education?" and looks into the idea of 'creative confidence' as an objective of design thinking education as taught at the d.schools in Potsdam and Stanford. Creative confidence refers to one's own trust in his creative problem solving abilities. Strengthening this trust is a main goal of d.school education. However, there have been only few efforts to develop the concept of creative confidence in design thinking on a deeper and measurable level. To substantiate this discussion, the team led by Birgit Jobst compares creative confidence with the concept of self-efficacy and discusses it in the context of d. school education.

The authors of the fourth chapter, **Steven P. Dow, Julie Fortuna, Dan Schwartz, Beth Altringer, Daniel L. Schwartz** and **Scott R. Klemmer**, take a closer look at prototyping dynamics in their chapter entitled "Prototyping Dynamics: Sharing Multiple Designs Improves Exploration, Group Rapport, and Results". Their assumption claims that prototypes ground group communication and facilitate decision making. However, overly investing in a single design idea can lead to fixation and impede the collaborative process. In a study, participants created advertisements individually and then met with a partner. In the *Share Multiple* condition, participants designed and shared three ads. In the *Share Best* condition, participants designed three ads and selected one to share. In the *Share One* condition, participants designed and shared one ad. Sharing multiple designs improved outcome, exploration, sharing, and group rapport. These participants integrated more of their partner's ideas into their own subsequent designs, explored a more divergent set of ideas, and provided more productive critiques of their partner's designs.

Chapter 5 is captioned with the title "Towards a Paradigm Shift in Education Practice: Developing Twenty-first Century Skills with Design Thinking". In this chapter, the authors **Christine Noweski, Andrea Scheer, Nadja Büttner, Julia von Thienen, Johannes Erdmann** and **Christoph Meinel** deal with the topic of science, business and social organizations, which all describe a strong need for a set of skills and competencies, often referred to as twenty-first century skills and competencies. For many young people, schools are the only place where such competencies and skills can be learned. Therefore, educational systems are coming

more and more under pressure to provide students with the social values and
attitudes as well as with the constructive experiences they need to benefit from
their opportunities and contribute actively to the new spaces of social life and work.
Contrary to this demand, the American as well as the German school system has a
strong focus on cognitive skills, acknowledging the new need, but not supporting it
in practice. The authors ask why this is so.

Authors **Adam Royalty, Lindsay Oishi** and **Bernard Roth** explore the
pathways to adaptive innovation in Chap. 6. Their article entitled "'I Use It Every
Day": Pathways to Adaptive Innovation' analyses why the demand for creative and
adaptive workers grows, and why universities strive to develop curricula that enable
innovation. A pedagogical approach from the field of engineering and design, often
called design thinking, is widely thought to foster creative ability; however, there is
little research on how graduates of design thinking programs develop and demonstrate creative skills or dispositions. This chapter proposes a new model for the
development of creative competence through design thinking education, and
investigates alumni outcomes from a graduate school of design thinking. The
authors explore potential mechanisms by which students develop these capacities
and foreshadow future analysis of obstacles to innovation in the workplace.

2.4 Part II: Design Thinking Research in the Context of Distributed Teams

Chapter 7, entitled "Tele-Board in Use: Applying a Digital Whiteboard System in
Different Situations and Setups," has been written by the authors **Raja Gumienny,
Lutz Gericke, Matthias Wenzel,** and **Christoph Meinel.** Tele-Board is a digital
whiteboard system that helps creative teams working together over geographical
and temporal distances. The nature of Tele-Board's synchronized setup allows
every connected partner from anywhere in the world to join in the action. Tele-
Board is rooted in traditional metaphors, which are easy to implement and come
naturally to the user. Additionally, it is possible to follow a common thread in the
development of ideas from their inception to conclusion. With the History Browser,
the path of creative development can be retraced, reiterated and resumed – from any
point in time – a huge benefit in ordering work and reaching conclusions. In this
article, the author team reports on several situations and setups in which Tele-Board
was used by different teams. They demonstrate how the software suite can be used
with various hardware setups and show how well the tools work in practical
application.

The authors of Chap. 8, **Greg L. Kress and Mark Schar**, take a closer look at
"Applied Teamology: The Impact of Cognitive Style Diversity on Problem
Reframing and Product Redesign Within Design Teams". Their chapter reflects
on the words of Professor Larry J. Leifer that "all design is redesign". As designers
collect information about a problem, they form a mental frame of the problem space

that is the scaffolding around which to build a solution. When presented with new information, successful designers can "reframe" the problem and the solution as part of a successful iterative cycle. The authors propose the Stanford Design Thinking Exercise (SDTE) as a measure of reframing behavior and design team effectiveness. They found that the SDTE is a robust frame for measurement of reframing change, in that it reports a range of reframing results across a participant population group. However, attempts to align the instrument with participant cognitive characteristics were unsuccessful, indicating that more work needs to be done to understand specific indicators of reframing.

Author **Jonathan Edelman** looks into "Qualitative Methods and Metrics for Assessing Wayfinding and Navigation in Engineering Design" in Chap. 9. He follows the assumption that designing can be viewed as a body of behaviors. Fundamental to several design behaviors is Path Determination. Path Determination describes the moments when designers choose what they will take up for development as well as how they experience their perceptual horizon. Their research suggests that there are two primary modes of Path Determination, Wayfinding and Navigation. Each of these has been correlated with different outcomes in redesign scenarios. Wayfinding correlates to making significant changes to an object, while navigation correlates to making incremental changes to an object. In this chapter, he presents a novel methodology for capturing and observing Wayfinding and Navigation behaviors, as well as several metrics for measuring these behaviors.

2.5   Part III: Design Thinking Research in the Context of Embedded Business Teams

In the tenth chapter, entitled "The Designer Identity, Identity Evolution, and Implications on Design Practice," authors **Lei Liu** and **Pamela Hinds** deal with the preliminary results of a study on designer identity, including what a designer identity is, how it evolves as a result of ongoing work-related interactions, and how it may influence design work practice. In their ethnographic research, they closely observed 12 in-house designers as they did their work in a major Chinese communication technology company. They found that designers identified with the design occupation in different yet non-mutual-exclusive ways, and that the way in which designers identified themselves influenced their creative thinking, brainstorming processes, and interactions with clients.

The authors **Martin Steinert, Hai Nugyen, Rebecca Currano and Larry Leifer** deliver insight at "AnalyzeD: A Virtual Design Observatory" in Chap. 11. This chapter describes the launch year activities of the analyzed project where the team led by Martin Steinert aimed to quantify engineering design behavior to such an extent that uses statistical algorithm to discover, describe and model fundamental design thinking behavior paradigm. This project is a joint research endeavor with the EPIC chair of Prof. Hasso Plattner at the Hasso Plattner Institute (HPI). As main result from the Stanford side, they were able to generate several proofs of concepts on gathering and analyzing design process data from various sources and in various data quality. Collaborating with a leading CAD software supplier, they were able to first extract every single engineer-system interaction and second, using genetic algorithms, they were able to statically identify patterns without an a priori model assumption.

In Chap. 12, entitled "When Research Meets Practice: Tangible Business Process Modeling at Work", the authors **Alexander Luebbe and Mathias Weske** are comprehensively deliver insight into a modeling approach they created and that is used by people in organizations to create and discuss business process models that represent their working procedures. This is an alternative to established approaches in which process modeling experts create business process models for the organization based on input from domain experts. They have changed this by empowering the domain experts to model their business processes themselves. This approach consists of a simple to use haptic toolset and the facilitation for its application. In the first stage of their research, they have shown that their approach, called tangible business process modeling (t.BPM), can be used to co-create process models with novice modelers. In a subsequent laboratory experiment, they found out that t.BPM is superior to interviews for process elicitation because people are more engaged with the modeling task and the result is better validated.

In Chap. 13, entitled "Towards a Shared Repository for Patterns in Virtual Team Collaboration," the authors **Thomas Kowark, Philipp Dobrigkeit**, and **Alexander Zeier** comprehensively analyze the platform they established and that provides 'out-of-the-box' monitoring capabilities for virtual team environments and enables the sharing and evaluation of recorded collaboration activities within a larger research community. Building on lessons learned from previous applications, they now present a refined and extended version of this platform. Its core feature is the possibility to share abstracted parts of the collected team collaboration networks with other users of the platform and, thus, broadens the basis for the validation of their influence on team performance. In this way, these network sequences might be raised in status from being just coincidently reoccurring collaboration behavior to collaboration patterns (or anti-patterns), whose occurrences are strong indicators for the possible success of teams.

Chapter 14 is titled "Adopting Design Practices for Programming". The authors **Bastian Steinert, Marcel Taeumel, Damien Cassou**, and **Robert Hirschfeld** are elaborating their article based on the assumption that developers are continuously designing their programs. In the same way, developers strive for simplicity and consistency in their constructions like practitioners in most design fields. A simple program design supports working on current and future development tasks. While many problems addressed by developers have characteristics similar to design problems, developers typically do not use principles and practices dedicated to such problems. In this chapter, the author team is reports on the adoption of design practices for programming. First, they propose a new concept for integrated pro-gramming environments that encourages developers to work with concrete representations of abstract thoughts within a flexible canvas. Second, they present continuous versioning as our approach to support the need for withdrawing changes during program design activities.

In Chap. 15, the authors **Gregor Gabrysiak, Holger Giese** and **Thomas Beyhl** give insight into their research topic, entitled "Virtual Multi-User Software Prototypes III". In design thinking and software engineering, prototypes play a crucial role in validating insights, needs and requirements. Still, the effort necessary to create these prototypes depends on multiple factors, such as the number of people involved with the design thinking project. Especially for multi-user software systems, the effort of creating validation artifacts is too high to be feasible for multiple iterations, thus, inhibiting design thinkers to inexpensively try different ideas early and often. To overcome this problem, the team led by Gregor Gabrysiak investigated the usability and feasibility of virtual prototypes – animated simulations of formal models which can be derived automatically without additional costs. This article reports on their advances during the 3 years of the design thinking research project concerned with Virtual Multi-User Software Prototypes.

In the last chapter of this book, Chap. 16, captioned "What Can Design Thinking Learn from Behavior Group Therapy" the authors **Julia von Thienen, Christine Noweski, Christoph Meinel, Sabine Lang, Claudia Nicolai** and **Andreas Bartz** look into the similarities of Behavior Group Therapy and Design Thinking. Some widely-used approaches in Behavior Group Therapy bear a striking resemblance to Design Thinking. They invoke almost identical process-models and share central maxims like "defer judgement" or "go for quantity". Heuristics for composing groups (mixed!) and preferred group sizes (4–6) are very much alike as well. Also, the roles ascribed to therapists are quite similar to that of Design Thinking coaches. Given these obvious analogies, it is most natural to ask what the two traditions can learn from one another – and why it is that they are so strikingly alike. This article by the author team led by Julia von Thienen ultimately hopes to inspire further investigations by giving examples of how Design Thinking may profit from taking a look at Behavior Group Therapy.

3 Summary

Understanding the evolution of innovation, and how to measure the performance of the design thinking teams behind this innovation, is the central motive behind the research work reported in this book. Challenged by these fundamental concerns, all the contributions in this volume report on different approaches and research efforts aimed toward obtaining deeper insights and a better understanding into how design thinking transpires. In highly creative ways, different experiments were conceived and undertaken with this goal in mind. The results achieved were analyzed and discussed to shed new light on the focus areas. We hope that you, as our reader, enjoy this discourse on design thinking and its multi-faceted impact. Besides looking forward to receiving your critical feedback, we also hope that when reading these reports you too will get caught up in the fun our research teams had in carrying out the work they are based on. All authors are very much interested in entering into or continuing a dialogue with you. We would be grateful if you could share your insights, impressions and ideas with us.

Understanding innovation, understanding how design thinking fosters innovations, that is the motivation for all the research work is reported in this book.

We are thankful to all who have contributed to the book. These are not only the authors but also Martin Steinert as well as untold helping hands from friends within the Stanford design and engineering community and the HPI. They all have successfully managed the program and various community building activities and workshops, all of which has contributed considerably to the success of the HPI Stanford Design Thinking Research Program.

We are particularly thankful to Sabine Lang for all her work in preparing this book and supporting its editors tremendously. We are very grateful for her various contributions that were a capital share into making this book happen.

We sincerely hope that you will enjoy and benefit from the content, format and intent of this book. We hope to instigate and contribute to scholarly debates and strongly welcome your feedback. You can contribute directly by submitting papers to the **"Electronic Colloquium on Design Thinking Research" (ecdtr)** which you can find here: http://ecdtr.hpi-web.de.

We invite you to visit this innovative platform of dynamic and rapid scholarly exchange about recent developments in design thinking research and to join in the dialog with us.

Part I
Design Thinking Research in the Context of Co-located Teams

Assessing d.learning: Capturing the Journey of Becoming a Design Thinker

Shelley Goldman, Maureen P. Carroll, Zandile Kabayadondo,
Leticia Britos Cavagnaro, Adam W. Royalty, Bernard Roth, Swee Hong Kwek,
and Jain Kim

Abstract The research explored the relationship of learning design thinking and assessing that progress. It addressed the questions: How can we understand what is learned in design thinking classes, and how assessments might contribute to that process in authentic ways? The study followed a reciprocal research and design methodology where basic research and the design of assessment solutions were ongoing, reciprocal, and related to each other in organic ways. We learned that the learning of design thinking dispositions and mindsets is an emergent journey—with various levels of sophistication, transformation, application, and integration. We introduce the concept of *mindshifts* to represent the developing and nascent epistemological viewpoints and instincts that are strengthened while becoming a design thinker. We

---

S. Goldman (✉)
Stanford University School of Education - Professor (Teaching) of Education and,
by courtesy, of Mechanical Engineering
e-mail: sgoldman@stanford.edu

M.P. Carroll
Stanford University School of Education - Research Staff

Z. Kabayadondo
Stanford University School of Education - PhD Candidate

L.B. Cavagnaro
Stanford University Technology Ventures Program - Associate Director

A.W. Royalty
Stanford University Hasso Plattner Institute of Design - Academic Staff

B. Roth
Stanford University School of Engineering - Professor and Hasso Plattner Institute of Design -
Academic Director

S.H. Kwek
Stanford University School of Education - Master's Degree Graduate

J. Kim
Stanford University School of Education - Master's Degree Graduate

H. Plattner et al. (eds.), *Design Thinking Research*, Understanding Innovation,
DOI 10.1007/978-3-642-31991-4_2, © Springer-Verlag Berlin Heidelberg 2012

review designs for tools that were based on the concept of mindshifts that include reflective and performance assessments and an assessment dashboard.

1 Introduction and Background

As students become design thinkers, they emerge with significant changes in their approaches to problem solving and to new challenges. They start to develop a sense of resiliency that enables them to think "outside the box." This research explored the experiences of students while they were enrolled in a university course at the Hasso Plattner Institute of Design at Stanford University and took up questions about the relationship between learning design thinking and assessing progress. To pursue the metaphor, assessment is often described as a "black box" activity, with students being tested and assessed outside of the actual learning process. Black and Wiliam (1998) describe how policies in the U.S. and in many other countries seem to treat the classroom as a black box, where certain *inputs* from the outside are fed into the box, and from which students emerge with certain competencies. Our goal was to understand, document, and design tools that could help assess what students were learning in design thinking courses. We sought to bring assessment "into the box" of the university design thinking courses as students were learning how to think "outside the box". We sought to understand the learning experience of the students in ways that would provide a reflection of progress made towards design thinking based on their authentic, course-based experiences. We hoped the work would have implications for both students and their course instructors. Students could better reflect on and have information about their evolving design learning. Instructors would be able to reflect on the progress of the individual and their design team, and use the information to reorganize teaching and curriculum to better meet the needs of students. The research addressed the following key questions: How can we understand what is learned in design thinking classes, and how might assessments contribute to that process in authentic and helpful ways?

The study followed a reciprocal research and design (RR&D) process and methodology (Alexander et al. 2010). In the RR&D model, basic research and the development of solutions are on-going, and are reciprocal and related to each other in organic ways. In this project, interim findings of studies of students in design thinking courses and activities were utilized as needs assessments for emergent assessment designs. This was done in such a way that after each round of data gathering and analysis, the preliminary results contributed to the creation of performance and reflection-based assessments. Design thinking aspects of the RR&D model influenced the research, which included interviews, observations, analysis/synthesis, and ideation to iterate successive prototypes of assessment tools for both students and their instructors. One such implement emerging from this process was the concept of a dashboard with tools for tracking how students are learning the art and practice of design thinking.

This paper chronicles the research and design process and its results. The research began by building on and adapting findings from a prior study we completed on assessment that occurred in K-12 educational settings and resulted in the development of a Design Thinking Assessment Rubric. That research resulted in a prototype of a rubric for documenting and assessing the skills and processes that students come in contact with while learning in the context of design thinking classrooms. The rubric allows the documentation of skills and processes learned such as interviewing, brainstorming and prototyping. That rubric provided a starting point for the research on assessment reported herein, which focused on graduate student design thinking classrooms at Stanford University's Hasso Plattner Institute of Design.

2 From Mindsets to Mindshifts

Through this study we have been able to rethink, redevelop, and redefine ideas about the assessment challenges related to design thinking. We were able to redefine the parameters and complexity of design thinking assessments and refine a rubric and its uses for students and their instructors. Ultimately, the study helped us gain further understanding of the different skills, stages, and mindsets of design thinking, their evolution among students, and their relationships to each other.

This last finding was most significant: we could develop assessments to capture the complexity and nested relationships of design thinking skills, processes and mindsets. One nuance of this complexity was to consider and capture, instead of static mindsets, *mindshifts* that result from the process of learning design thinking came into being. Since we consider the learning of design thinking dispositions as an emergent journey—with various levels of sophistication, transformation, application, and integration—we use the term mindshifts to represent the active shifts that students are making. Mindshifts are the developing and often nascent epistemological viewpoints and instincts that are strengthened (or instilled) and made observable through change in a learner's orientations and actions as a design thinker. Mindshifts are significant in the design thinking learning arc. For many students they are the re-synthesis and reorientation of their worldviews, routes, and propensities in problem solving. They require new categories and sensibilities. Even as students are making changes to their categories or worldviews, they hold on to and integrate old and new ways of thinking and acting. There is a long history of ideas about how conceptual shifts can be understood. Wallace (1956) examined how changes in cultural thinking and worldviews were possible. Sometimes the shifts were violent and intense, and at other times they were more gradual, with a back and forth between old and new views of the world and how to act in it. Mindshifts are not always at the cultural level, and can occur at more local levels. Carol Dweck's work (2007) on fixed and growth mindsets shows how growth mindsets can be learned—hard work and a propensity towards problem solving and action can lead to great learning and to productivity. Studies in physics education also show how students' naïve ideas about the physical world could be

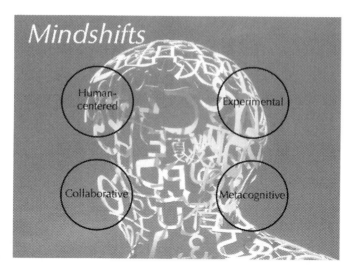

Fig. 1 Mindshifts

reorganized through instruction (Smith et al. 1993). We think of mindshifts in a similar vein. Design thinking students accomplish new mindshifts as they learn and become experienced. Design thinking mindshifts differ from design skills and processes. They often seem synthetic to other skills and processes, and are the superstructure of a learned design thinker. They are difficult to observe, but it is possible to uncover and identify them. These ideas about mindshifts led to a framework for reflective assessment activities, a performance-based rubric for skills and processes, and an assessment dashboard for managing information about what is learned.

As such, the conception of mindshifts challenged our assumptions about how to assess design thinking. We initially focused primarily on process and skill development and on how to assess these, but later prioritized the identification of mindshifts, which we defined as epistemological viewpoints that are evident by a change in one's orientation as a design thinker (Fig. 1). This understanding led us to a deeper investigation of the transitional nature of what it means to adopt the orientation of a developing design thinker.

Of the conceptual shifts we identified, we focused on four key mindshifts: human-centered; experimental; collaborative; and metacognitive.

2.1 Human-Centered

The human-centered mindshift is characterized by a central focus on empathy for others. We focused our work specifically on assessments that help make visible and document the development of students' human-centered mindsets. When students are developing a human-centered mindset, they begin to move beyond egocentric views of the world and no longer design based on their own needs, desires, experiences or preferences. Becoming human-centered is a fluid and dynamic process where students actively seek solutions to problems that meet the needs of others who might benefit from their innovation or design. Each engagement takes students toward being more human-centered and can help them see, consider, interact with, and have empathy for others. A human-centered mindset is an integral, necessary and distinguishing element of design thinking. While we see that students can accomplish human-centered mindshifts, we suggest that the tasks developed and researched can be perform as near transfer tasks for examining students' development of a "human centered" mindset.

2.2 Experimental

The experimental mindshift is characterized by a realization that everything may be considered a prototype. Having an experimental stance changes one's approach to problem solving by allowing one to do, make, and visualize as integral parts of thinking and of the evolving ideas.

2.3 Collaborative

The collaborative mindshift is characterized by a belief that collaboration is a key component of problem solving, and that radical collaborations undergird transformative innovation.

2.4 Metacognitive

The metacognitive mindshift is characterized by an awareness that it is essential to be aware of where one is in the design thinking process in order to agilely respond to changing parameters of a problem.

3 Needs

It is critical that students master both the knowledge of core subjects and the critical skills necessary for readiness in the innovation economy of the twenty-first century. Twenty-first century skills include critical thinking and problem solving, communication, collaboration, and creativity and innovation (Partnership for 21st Century Skills 2008). Content learning, life skills, innovation, and fluency with technologies such as media and web functionalities are also included. These skills are considered essential for developing a participatory, innovative, and technologically sophisticated culture; and contribute to the personal, educational, professional, and civic lives of learners (Jenkins 2006; Jenkins et al. 2006). Design thinking, with its focus on problem solving and creative confidence, is nested within this view of cultural shift. Such new learning goals create a critical need to understand how design thinking is integrated into classroom learning environments.

A need for new metrics and assessment methodologies accompanies these new ways of learning, as educators become increasingly aware that measures of student progress need to align with the new skills required for school, entry to the workforce, and life (COSEPUP 2006). The US National Science Foundation Task Force on Cyberlearning (2008) described this need as follows:

> Despite the revolutions wrought by technology in medicine, engineering, communications, and many other fields, the classrooms, textbooks, and lectures of today are little different than those of our parents. Yet today's students use computers, mobile telephones, and other portable technical devices regularly for almost every form of communication except learning. The time is now—if not long overdue—for radical rethinking of learning and of the metrics for success.

President Obama has also recognized the critical need to measure twenty-first century skills, calling on the nation's governors to:

> ... develop standards and assessments that don't simply measure whether students can fill in a bubble on a test, but whether they possess 21st century skills like problem-solving and critical thinking and entrepreneurship and creativity (Obama 2009).

Most of the assessment done today is after the fact and designed to indicate only whether students have learned. Not enough is done to collect and aggregate data about student learning that makes the information valuable to and accessible by stakeholders who might support continuous improvement and innovation. Very few assessments are authentic to classroom-supported learning activities or to the application of what is learned out of school situations (Baker 2010). Generally, assessments do not give students access to, or active participation in, their own learning progressions. Silva (2008) describes the creation of effective measures of twenty-first century skills, and the need for better and more comprehensive assessments that measure both the ability to think creatively and to evaluate information, and student's mastery of core content or basic skills and knowledge. These notions of complex, holistic and performance-based measures relating to innovation learning were central to this project's research focus and the development of assessment tools that could begin to capture the complexity of design learning in progress.

4 Theoretical Perspectives

The five key phases of the design thinking process are empathize, define, ideate, prototype and test (Hasso Plattner Institute of Design 2007). Together, they provide each learner with a relevant, socially situated, complex problem-solving environment in which to generate solutions. While learning to become design thinkers, students focus on defining the parameters of a problem, identifying needs, dealing with varying levels of ambiguity, actively solving problems and making connections between their lives inside and outside of school. As researchers of design thinking who are also concerned with learning, we were guided by a theoretical stance that is based in experiential and sociocognitive views of learning. As Vygotsky (1934/ 1976) described, opportunities to interact with others who have varying degrees of expertise in a social environment become crucial to learning. The human-centered focus of design thinking and the deep and radical collaborations that define the process provide a deeply social process for learning. Design thinking is an approach toward learning that encompasses active problem solving by engaging with (Dewey 1916), and changing, the world. Language is central to this view, as we communicate and engage in dialogue with others (Bakhtin 1981). Learning, as Freire and Macedo (1987) describe it, demands a critical way of comprehending and of realizing the reading of the word and that of the world, the reading of text and of context. Design thinking supports this Freirean notion of impactful change which focuses on learning as a process where knowledge is presented to us, then shaped through understanding, discussion and reflection and, ultimately, action on behalf of others. Furthermore, design thinking is also informed by Papert's (1980) view of constructionism, which describes how ideas get formed and transformed when learners are involved in making tangible objects that are expressed through different media, actualized in particular contexts (Kafai and Resnick 2000; Todd 1999). This transformation, which Haskell (1985) describes as the acquisition of a know-how skill and new technologies, inspires the subject with an irresistible urge to change the lives of strangers who those technologies and skills can impact.

Design thinking provides a robust scaffold for divergent problem solving, as it engenders a sense of creative confidence that is both resilient and highly optimistic. The need for this kind of approach is timely. According to the Carnegie Foundation Commission on Mathematics and Science (2009), the United States needs an educated young citizenry with the capacity to contribute to and gain from the country's future productivity, understand policy choices, and participate in building a sustainable future. Likewise, research has indicated that many first-year college students need further development in critical thinking and problem solving (Lundell et al. 2004). The need for knowledge and skills from science, technology, engineering, and mathematics are considered crucial to virtually every endeavor of individual and community life. And, in a time of economic uncertainty, design thinking has the potential to address the demands of developing, recruiting and retaining top students and agents of innovation.

Design thinking, as a mode of inquiry that puts "doing" and "innovating" at the center of problem-solving, promises to address future needs of the globe. It has the potential to engage students in ways that are inclusive of their diversity, makes school learning relevant and real, pressing local and global issues which can enhance one's motivation to learn. It creates a "third space" (Gutiérrez 2008), an interdisciplinary space where students respond to design challenges with a clearly articulated sense of their confidence and agency and, more specifically, of their identities as change agents.

At the university level, the commitments of both Stanford University and the University of Potsdam to design thinking education and research are leading the way to better understandings of what is taking place in design thinking education and creating tools that help capture and improve the process at the university level (Meinel et al. 2011; this volume). Design thinking education at any level has the potential to impact learning skills such as working in groups, following a process, defining problems, and creating solutions (Barron 2006). Some research in the K-12 arena shows that design-based learning can be productive even before the university years (Carroll et al. 2010; Hmelo et al. 2000; Goldman 2002). Early work in this domain has indicated the potential for design to contribute to young people's metacognitive (Kolodner et al. 2000 and 2003) and social learning (Cognition and Technology Group at Vanderbilt 1997) as well as in specific subject areas (Goldman et al. 1998; Middleton and Corbett 1998). Vande Zande (2007) characterizes design thinking as a means of creative problem-solving that relates thought and action directly and dynamically. The methodologies of design thinking and rapid prototyping play important roles in developing transformative advances in learning and teaching (Cobb et al. 2003; Design-Based Research Collective 2003). Mimicking this orientation towards creative problem-solving, radical collaborations, and learning through design in our subject matter, a design thinking research methodology, the team was able to orient itself to the complex problem of assessing design thinking.

Two types of studies defined the research agenda and process. The first set of studies took an exploratory approach, using observations and interview methods to generate a detailed understanding of students' design learning just after they were ensconced in courses. The second series of studies were aimed at testing the assessment tool prototypes that were designed in conjunction with our exploratory study results. The two types of studies were reciprocal, taking our team back and forth between them to refine findings in an iterative mode of understanding learning and assessment. Here, it was our intention and practice for findings based on exploratory studies to contribute to a prototype assessment tool. When analyzing results of testing on the tools, we then poured those results back into more refined or nuanced understandings of the original basic research findings.

5 Research Methods and Analysis

Qualitative methodologies (Bogdan and Biklen 1992) were employed in data collection and analysis. The exploratory studies took a student-centered, emic approach and were conducted with students who were in Stanford design thinking courses. We conducted observations and interviewed students who were enrolled in two of Stanford's Hasso Plattner Institute of Design (d.school) classes: *Design Thinking Bootcamp*; and *Design Garage*. The observations took place during the Fall and Winter academic quarters. Observations, which focused on the full class and on smaller project teams, were recorded in field note records. At the end of the course, observations were followed with open-ended interviews with student participants. Interview questions covered topics relating to the overall experience in the class. Some asked them to reflect on, and describe, when and what they thought they were learning. Sample questions from the interview included:

What has been your biggest struggle?
What has been your biggest surprise?
Please draw a visual representation of design thinking as you understand it.
Tell us about an experience of boot camp that made an impression on you.
What do you think will stay with you after the class is completed?

With the observations and interviews completed, analyses were conducted to identify the kinds of learning experiences students were having and to determine how they met the goals set for them by their instructors. Analysis techniques were qualitative and included open coding of the field notes and interviews, searches for important language, events, and "learnings" that students described, and places where students expressed uncertainty about what or whether they were learning. We searched for times, places and events where feedback came into play for the student: how they understood the feedback they were getting; when they reported they wanted feedback that they did not receive; and what they thought that meant for them as budding design thinkers. This protocol helped us identify opportunities for feedback or reflection on performance; expressions of learning; and documentation of learning. For example, in one class we observed students asking for confirmation that they were "getting this ideation process correct" during class, and being told, "Don't worry about whether or not you are getting it at this point in time. Just give it a try!" Students were not always reassured at those times, but forged onward. For a few students, we could actually trace how those classroom moments played out in their own thinking or in the feedback they received on class projects. This is an example of a series of assessment opportunities that we could identify and try to understand.

With the preliminary exploratory study findings at hand, we began to design a set of assessment tools. We conducted formative user-tests with these tools to improve them iteratively and to further refine insights from our exploratory findings. Through user-testing feedback on the assessment tools and exploratory observations, we were able solve our ongoing dilemma about how difficult it seemed to assess design thinking mindsets. The studies indicated that mindset development was difficult to observe and document during classroom activities even though they seemed present when you talked to students about their experiences and design projects. The cycling back-and-forth among the studies helped us see that the assessment rubric was not an appropriate vehicle for assessing the students' evolving mindshifts. We realized mindshifts could best be assessed and benchmarked through specialty performance assessment tasks and prompts.

In order to develop performance assessment tasks that could measure the development and presence of mindsets, we observed and interviewed school children, university students, and instructors and experienced design thinkers. We focused our studies on one mindset—being human centered. We developed and researched two tasks—one task for student groups and one task for individuals—which would show how students approached and worked through challenges. We could thus explore how students were exhibiting a human-centered mindset and document the level of sophistication at which this occurred. We administered the tasks to middle school, undergraduate, graduate students, and design thinking instructors. This enabled us to examine and hopefully distinguish among performances of novice, intermediary and sophisticated design thinkers. As we hoped, an analysis of stark differences between the novices and experts revealed a complex pattern of development, and indicated that mindsets might be more in flux than "set". A way to conceive of and describe some of the common evolving behavioral moves for novice, intermediate and solidified mindsets was therefore the starting point for the concept of evolving *mindshifts* and a new round of development and validation of the reflective assessment rubric and the performance tasks.

6 Assessment Tool Prototypes

Introductions to the assessment tools that were prototyped and studied follow. Each introduction includes a basic description of the tool. We conducted observations and interviewed students who were enrolled at Stanford's Hasso Plattner Institute of Design (d.school) classes: *Design Thinking Bootcamp* and *Design Garage*, and also students from Stanford's School of Education *Learning, Design & Technology* program. With insights from these research experiences in place, we began to iterate a set of tools that would measure both mindshift (the active process of developing a mindset) and process. We focused on the creation of performance-based tools and reflective tools that assessed how one would approach both problem definition and problem solving. These would capture the essential shifts that occur when one develops as a design thinker. Our implements include (1) the *Reflective Assessment Rubric*; (2) the team-based *Windaloobah Experiment* task; (3) the *Designing Twenty-first Century Learning Spaces* Task; and (4), the *Assessment Dashboard* concept.

6.1 The Reflective Assessment Rubric

With an initial prototype of a design thinking assessment rubric developed in a prior research project, we used the iterative design thinking approach to continue evolving the rubric. Our activities centered on: interviews, observations, analysis/synthesis, ideation, prototyping and testing, as focal points of the design process. The learning from each type of prototype informed successive versions of the assessment rubric. Figure 2 illustrates a sample page from an advanced prototype of the reflective assessment rubric.

Observations of expert and novice designers led to a matrix of skills, processes and mindsets related to design thinking. When we considered how they were noticed and could be documented, we realized that different kinds of assessments were needed for assessing skills, processes, and mindsets. Some were nested; others

Skills & Processes

Mindshift (HC, E, C, MC)	Process (E,D,I,P,T)	Skill	Level 1	Level 2	Level 3
2	Human-centeredness	E,T	Interviewing-questions	Asks topic-based and closed questions	Asks a mixture of topic-based and open questions, asks some follow-up questions	Asks questions to elicit stories; Makes use of prompts such as "Why?" and "Can you tell me more?"
9	Human-centeredness	P	Prototyping -point of view	Designs for oneself; is unable to recognize the purpose of designing for someone else	Recognizes the difference between designing for oneself and designing for someone else	Recognizes and enacts designing for someone else
16	Experimental, Human-centeredness	T	Prototyping-presenting	Sells prototype rather than tests prototype; Unwilling to let tester explore, instead, relies on explaining the prototype	Able to test a prototype and see value of feedback; Listens to tester	Able to strategically test a prototype to garner feedback by allowing tester to freely explore
17	Experimental	T	Prototyping	Unable to use prototype to collect data	Able to test a prototype with a user and collect 2-3 feedback insights	Able to test a prototype with a user and collect 5-10 feedback insights
19	Experimental	P	Prototyping	Is attached to first prototype	Able to let go of first prototype	Able to easily let go of first prototype
22	Experimental	P,T	Prototyping	Does not understand the value of missed understandings or unexpected outcomes	Understands the value of missed understandings or unexpected outcomes	Understands the value of missed understandings or unexpected outcomes and uses this information to iterate
24	Experimental	P, T	Persistence	Gives up easily; easily frustrated;	Willing to attempt a task again after a failure;	Embraces failure; has a positive and energetic attitude
27	Experimental	I, P	Risk-taking	Is uncomfortable trying things he/she has not done before	Is willing to try new things; expresses some reluctance	Is totally open to trying new things; expresses excitement about the unknown; Sees risk-taking as a valuable way to learn new things
29	Metacognitive, Collaborative	I	Brainstorming	Unable to headline when expressing ideas; gives extended explanations	Begins to use headlining to streamline the expression of ideas	Able to fluidly and consistently headline ideas
30	Collaborative	I, P	Brainstorming	Does not build on the ideas of others	Sometimes builds on the ideas of others	Consistently is inspired by others' ideas and often builds on them
35	Experimental	I, P	Storytelling	Begins to use new tools for expressing ideas, for example, podcasts, murals, music	Is comfortable exploring how to use new tools for expressing ideas, for example, podcasts, murals, music	Seeks out new tools for expressing ideas, for example, podcasts, murals, music, and makes strategic use of them to tell a story
35	Collaborative	I, P	Collaboration - Contribution	Makes a limited contribution to group efforts	Consistently contributes to group efforts	Contributes to and leads group efforts

Fig. 2 Sample page from the reflective assessment rubric

were umbrella skills. Some were discrete; others were bundled. Some could be documented and much repetition and practice put them into place; others could be accomplished once and thenceforth seem to take root. Some design thinking skills overlapped with skills from other domains, for example, interviewing, observation, and persistence skills. Other skills, such as those related to human-centeredness, ideation, and prototyping were more exclusive to design thinking. We reconsidered the rubric's categories and chose specific types of actions to be included that best exemplified design thinking. We also separated mindshifts from rubric assessment, and turned attention to *mindshifts*.

To date, the rubric consists of mindshifts, skills, and three levels of expertise. It also indicates where a specific skill is exhibited in the process and where skills tend to overlap and interact with mindshifts. The identified skills include the following:

- Interviewing
- Prototyping
- Synthesis
- Persistence
- Resilience
- Adaptability
- Risk-taking

- Brainstorming
- Bias Towards Action
- Storytelling
- Process Vocabulary
- Collaboration

A specific skill may present itself in more than one phase of the process. For example, one might interview during both the empathy and testing phase of the design thinking process, which will, consequently, reflect different components of the skill. Students of design thinking learn at different rates, and the three levels are approximate representations of behaviors and actions for novice, intermediate and skilled design thinkers. When analyzing data, we also found that students exhibited a range of skills that were indicative of their abilities as developing design thinkers. Levels of expertise varied for each phase of the process. While working on the rubric we considered who would best be in a position to assess developing skills and mastery of processes. While many items called for assessment by an instructor, we also identified many items on the rubric that could be documented by students. For example, students could track their work and progress relating to topics such as persistence, self-regulation, and development of meta skills. The rubric is available electronically and we continue to validate it based on feedback from students and instructors.

6.2   The Windaloobah Experiment Task

Once we discovered that the rubric was an inappropriate instrument for assessing *mindshifts*, we transferred our efforts to the design of performance-based assessment tasks.

We created an assessment task, *The Windaloobah Experiment*, to gauge students' understanding of the design thinking mindsets. The task was designed to give students, working in teams, opportunities and resources for exhibiting empathy and human-centeredness. It was administered to two classes of middle school students and two groups of master's level graduate students. In the middle school, one class had recently completed an integrated mathematics and design thinking challenge. The second class had not yet been introduced to design thinking. The graduate students were diverse in experience, ranging from none to a student who was in the d.school class *Design Thinking Bootcamp*.

The Windaloobah Experiment was structured as an hour long design challenge. The students were asked to work in teams to design a *Windaloobah*, however, the definition of a *Windaloobah* (a made up word) was deliberately ambiguous. Students watched a brief introductory video. Figure 3 contains the video script from the video.

After watching the video, the students were given information packets with profiles of people who would be part of the community embarking on a journey on the Windaloobah. The profiles consisted of a photograph of each community

Welcome to the *Windaloobah* Experiment. It is the year 2125. A group of brave and adventurous explorers have joined the *Windaloobah* experiment. For the next 15 months they have no permanent home. Your job is to design a *Windaloobah*.

What is a *Windaloobah*?
- A *Windaloobah* must have room for 5 people.
- A *Windaloobah* must travel through water, skies, snow, it must be able to move through cities. It must be ready for unexpected adventures.
- A *Windaloobah* is a home away from home.

Fig. 3 *The Windaloobah Experiment* video script

Fig. 4 *Windaloobah* community member profile

member, and text describing his or her interests, needs and challenges (See example profile in Fig. 4). The Windaloobah community included a grandmother, a 3-year-old boy, a shy 13-year-old girl, a 23-year old musician, and a 19-year-old videographer. The students were also given an envelope with a travel itinerary describing the conditions the travelers would encounter, and a set of pictures (a kite, a sign with 5 mph on it, and a squirrel.) The students had to ascertain what they might do with ambiguous information. Each group was also given a cardboard box filled with an assortment of prototyping materials.

The Windaloobah Experiment task enabled us to observe and document human-centeredness as groups of students engaged a design challenge. With the task, we could discriminate human-centeredness, separating out how human-centered design thinkers approached the task differently. Students who had even an introductory level of design thinking showed more human-centered behaviors and actions than those who were completely inexperienced.

6.3   The Designing Twenty-First Century Learning Spaces Task

The research team was also interested in designing a reflection-based task for individuals in order to gain further insight into human-centered mindshifts. Students were invited to a website, *Designing Twenty-first Century Learning Spaces,* where they were asked to provide input (Fig. 5). They were asked to watch a video that consisted of two components of exploring the issue. The first part of the video focused on a classroom filled with an array of technology materials. There were no students in the room. The second part of the video featured a story about a teacher, Nate, in an urban classroom who was concerned about meeting the needs of his students. At the end of the video, the research subjects were directed to a survey and asked to share their thoughts. The survey consisted of two questions:

In your opinion, what is the most important aspect to consider in the design of the classroom of the future?
What would you do if you had to design a classroom for Nate?

The survey also included a question regarding the responder's experiences with and exposure to the design thinking process. Preliminary data analysis indicated that students with no design thinking experience showed a baseline of little to no human-centeredness. If they did orient to the Nate's stated needs, it was limited to a level of "noticing" which did not extend to design solutions. At present, the task needs guidelines for scoring responses, and in future work we would like to validate the task with a large number of participants.

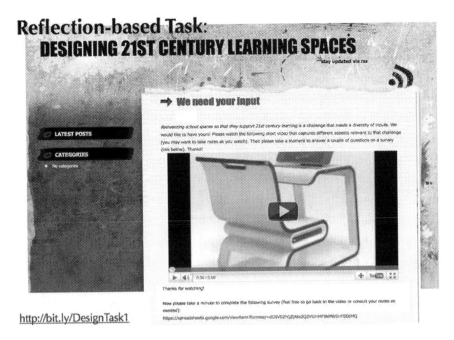

http://bit.ly/DesignTask1

Fig. 5 Designing twenty-first century learning spaces

6.4 The Assessment Dashboard

From the time we first prototyped the assessment rubric, we were aware that instructors and students would want a way to keep track of design learning over time. Design thinking is a complex of interwoven and interdependent skills, processes and mindshifts. Even students in the same course and design challenge are likely to develop different skills, capabilities and mindsets. Thus, in each study we completed, we looked for ways to think about and organize the assessment process and the snapshot on learning that it generates. We initially had the idea of a longitudinal portfolio that would follow students and their learning over time. At the time all of our effort was focused on the assessment rubric. Once we moved to a notion of varied palette of assessment types, we had to rethink how to organize assessment information so it was useful and transferable to stakeholders.

We tried different ways to organize mindshifts with skills and practices. We decided to treat mindshifts as reflective tasks where student-simulated approaches to design thinking action would help reveal the more hidden propensities of mindshifts. In order to allow students and teachers to effectively visualize these relationships, make explicit connections with their projects, and document a student's progress in developing as a design thinker, our work turned to designing an online dashboard. The dashboard was conceived to consider many of the needs

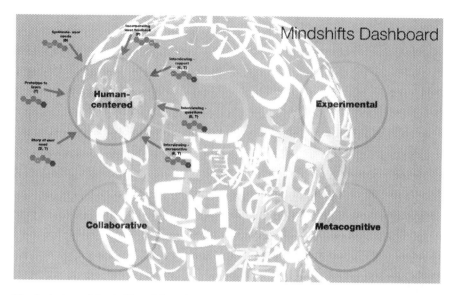

Fig. 6 A view of the dashboard from the mindset view

we were discovering. Figures 6 and 7 illustrate the framework for a dashboard that might meet the needs of individual students, students working in teams, and instructors who must weigh in on learning for both instances. This framework would allow stakeholders to document the different aspects of design learning and would serve as an assessment environment that traveled with the student over time. The dashboard would archive media related to design challenges and benchmarks in design learning, helping trace the changing levels of sophistication as a design thinker grew over time.

To date, we have developed a first prototype of the dashboard and will be moving towards user feedback and further iterations.

The dashboard is designed to enable instructors and learners to:

- Visualize the skills that are relevant to stages of the design thinking process
- Visualize the skills and process stages that contribute to each mindshift (Fig. 6)
- Build a portfolio (Fig. 6), selecting those stages and skills focused on in a given project. Developing mindshifts are included.
- Assess and view the performance of individuals and teams for different projects, progress over time, and for each of the mindshifts (Fig. 7).

The dashboard will also allow the research team to prototype and refine the assessment of the skills that are included in the rubric by tracking its use by teachers. At the moment the dashboard is in the user feedback phase.

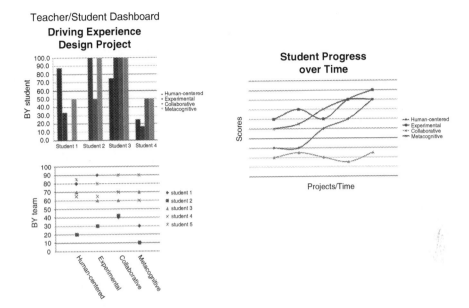

Fig. 7 Visualizing progress over design projects

7 Summary and Discussion

Through a series of reciprocal research and design studies we explored critical elements of learning the design thinking process and possible ways to document them. Our focus on the development of performance-based assessment tasks, on reflection-based assessment tasks, and on the further development and refinement of an assessment rubric made for rich learning. The results of the project helped enrich our assessment tools and our conceptualization of design learning. Firstly, we understood the complex relationship and differences between design thinking skills and processes, and the development of design thinking mindsets. These differences had implications for the assessment tools. Secondly, we conceived *mindshifts*, which express the developmental journey towards mindsets that are so important when a person is learning to be a design thinker.

Mindshifts are epistemological viewpoints in flux. They are part of the process of becoming design thinkers. We were able to identify four central mindshifts—becoming human-centered, experimental, collaborative, and metacognitive—each of which are representative of different aspects of one's development as a design thinker. Our perspective change to *shifts* gave us new insights into the most significant ways that one's behavior and orientation change as one grows as a design thinker. A *mindshift*, when finally set, is an underlying structure for thinking and acting as a design thinker—it is the deep structure and world-view that is learned in conjunction with, and supports, the development of design skills. For example, one might be a good prototype developer, but someone with an experimental *mindshift* understands the value of prototyping and seems to have instincts for bringing prototyping into the process early and often. One might be a skilled interviewer, but being an empathetic interviewer—the manifestation of the human-centered mindshift—is essential to being a design thinker.

Prior to understanding the mindsets as *shifts* in perspectives and development of predilections instincts, we focused on the skills and processes of design thinking that were observable and documented based on the performance of tasks and activities. However, after this change in orientation, we focused our work specifically on assessments that help make visible and document the development of students' human-centered mindshift. When students are developing a human-centered mindset, they move towards solutions that resonate with the needs and lives of others, marking a sharp contrast to egocentric views of the world that are characteristic of a failure to adopt human-centeredness. Mindset development is a fluid and dynamic process, yet for some students, developing a mindset might seem like a "switch has been flipped from off to on". For others, mindshifts come in more complex and incremental ways. Each step a student takes toward being more human-centered can help him or her see, consider, interact with, and have empathy for others. Our research team and other design thinking educators took for granted that some or all students would gain mindsets, but during this project we were actually able to witness students reaching benchmark points in the evolution of these mindsets. Such students were in the process of mindshifts. We saw that students might have no inclination toward human-centeredness, they might tend to act in human-centered ways intermittently or at convenient stages of the design process; or they might consistently orient themselves in human-centered ways no matter where they were in the design process. Our reflective tasks aided us in seeing differences in those who were "mind-shifting" and those who were not. Likewise, our research with students who had little or no experience with design thinking courses showed a stark contrast with those who were taking courses and those who were already considered more professional design thinkers.

We were able to expand and refine design thinking skills that were included in the rubric and identify actions and behaviors that might be part of the documentation process. Through the RR&D process, we identified topics for further research and design of assessment tools for the design thinking classroom. The following questions will be the focus of future work:

1. How effectively do the performance-based assessment rubric and reflection-based tasks document the skills and *mindshifts* inherent in the design thinking process? How useful are they as tools for both students and instructors?
2. Are the tools useful in a variety of educational settings as indicators of design learning?
3. Since the tools were developed to assess, document, and help individuals and their instructors in design thinking education, what needs to be accounted for in developing assessments of groups who are learning design thinking? Even with a demonstrated strong need for such tools, are these assessments practical and usable given the structure of courses?

We made progress toward our goal of providing a detailed reflective framework for students and teachers to document, reflect on, and confirm their engagements and progress in relation to a diverse range of design thinking activities. The framework is accompanied by a set of tools for tracking and certifying students' progress in relation to design thinking skills, processes and mindsets. In addition, the tools can provide a shared resource for design thinking teams and instructors. It is essential to develop a comprehensive understanding of design thinking from theoretical and applied vantage points. It is our hope that our research will contribute to this understanding.

Acknowledgments We would like to thank the students and instructors who participated in our studies. They have made it possible for us to better understand what it means to become a design thinker and how assessment tools could add value to design learning. A grant from the Hasso Plattner Design Thinking Research Program made this work possible. Findings and opinions presented are those of the authors and do not represent the HPDTRP.

References

Alexander A, Blair KP, Goldman S, Jimenez O, Nakaue M, Pea R, Russell A (2010) Go Math! How research anchors new mobile learning environments. In: Proceedings of the sixth international IEEE conference on wireless, mobile, and ubiquitous technologies in education (WMUTE). Kaohsiung, pp 57–64

Baker E (2010) What probably works in alternative assessment. CRESST Report 772. National Center for Research on Evaluation, Standards, and Student Testing, Los Angeles

Bakhtin MM [1930s] (1981) The dialogic imagination: four essays. In: Holquist M (ed) (Trans: Emerson C, Holquist M). University of Texas Press, Austin/London

Barron B (2006) Interest and self-sustained learning as catalysts of development: learning ecology perspective. Hum Dev 49(4):193–224

Black P, Wiliam D (1998) Inside the black box: raising standards through classroom assessment. Phi Delta Kappan, 80(2):139–148, October 1998

Bogdan R, Biklen S (1992) Qualitative research for education: an introduction to theory and methods. Allyn & Bacon, Boston

Carnegie Foundation Commission on Mathematics and Science (2009) Excellence and equity in mathematics and science to transform education. http://opportunityequation.org/report. Accessed 29 Jan 2011

Carroll M, Goldman S, Britos L, Koh J, Royalty A, Hornstein M (2010) Destination, imagination, and the fires within: design thinking in a middle school classroom. Int J Art Design Educ 29 (1):37–53

Cobb P, DiSessa A, Lehrer R, Scauble L (2003) Design experiments in educational research. Educ Res 21(1):9–13

Cognition and Technology Group at Vanderbilt (1997) The Jasper Project: Lessons in Curriculum, Instruction, Assessment, and Professional Development. Lawrence Erlbaum Associates, Inc., Mahwah, NJ

Committee on Science, Engineering and Public Policy (COSEPUP) (2006) Beyond bias and barriers: fulfilling the potential of women in academic science and engineering. National Science Foundation, Arlington

National Science Foundation Task Force on Cyberlearning (2008) Fostering learning in the networked world: the cyberlearning opportunity and challenge. http://www.nsf.gov/pubs/2008/

nsf08204/nsf08204.pdf http://www.nsf.gov/pubs/2008/nsf08204/nsf08204.pdf. Accessed 20 Dec 2010

Design-Based Research Collective (2003) Design-based research: an emerging paradigm for educational inquiry. Educ Res 32(1):5–8

Dewey J (1916) Democracy and education. Macmillan, New York

Dweck C (2007) How not to talk to your kids: the inverse power of praise. New York Magazine. http://nymag.com/news/features/27840/. Accessed 8 Nov 2011

Freire P, Macedo D (1987) Literacy: reading the word and the world. Bergin and Garvey, South Hadley

Goldman S (2002) Instructional design: learning through design. In: Guthrie J (ed) Encyclopedia of education, 2nd edn. Macmillan Reference USA, New York, pp 1163–1169

Goldman S, Knudsen J, Latvala M (1998) Engaging middle schoolers in and through real-world mathematics. In: Leutzinger L (ed) Mathematics in the middle. National Council of Teachers of Mathematics, Reston, pp 129–140

Gutiérrez KD (2008) Developing a sociocritical literacy in the third space. Read Res Quart 43 (2):148–164

Haskell TL (1985) Capitalism and the origins of the humanitarian sensibility, parts 1 and 2. Am Hist Rev. 90(3, 4):339–361, 547–566

Hasso Plattner Institute of Design at Stanford (2007) Design thinking process. Stanford University, Palo Alto

Hmelo C, Holton D, Kolodner J (2000) Designing to learn about complex systems. JLearn Sci 9 (3):247–298

Jenkins H (2006) Convergence culture: where old and new media collide. New York University Press, New York

Jenkins H, Clinton K, Purushotma R, Robinson AJ, Weigel M (2006) Confronting the challenges of participatory culture: media education for the 21st century. The MacArthur Foundation, Chicago

Kafai Y, Resnick M (eds) (2000) Constructionism in practice: designing, thinking, and learning in a digital world. Lawrence Erlbaum Associates, Mahwah

Kolodner J, Gray JT, Fasse BB (2000) Promoting transfer through case-based reasoning: rituals and practices in learning by DesignTM classrooms. Cogn Sci Q 1:183–232

Kolodner JL, Camp PJ, Crismond D, Fasse B, Gray J, Holbrook J, Putambeckar S, & Ryan M (2003) Problem-based learning meets case-based reasoning in the middle-school science classroom: Putting Learning By Design Into Practice. The Journal of the Learning Sciences, 12 (4): 495–547

Lundell DB, Higbee JL, Hipp S, Copeland RE (2004) Building bridges for access and success from high school to college: proceedings of the metropolitan higher education consortium's developmental education initiative. Center for Research on Developmental Education and Urban Literacy, University of Minnesota, Minneapolis

Meinel C, Leifer L, Plattner H (eds) (2011) Design thinking: understand–improve–apply. Springer, Heidelberg

Middleton JA, Corbett R (1998) Sixth-grade students' conceptions of stability in engineering contexts. In: Lehrer R, Chazan D (eds) Designing learning environments for developing understanding of geometry and space. Lawrence Erlbaum Associates, Mahwah, pp 249–266

Obama B (2009) Address to the Hispanic Chamber of Commerce. U.S. Department of Education, Assessment: measure what matters. http://www.ed.gov/technology/netp-2010/assessment-measure-what-matters. Accessed 20 Dec 2010

Papert S (1980) Mindstorms. Children, computers and powerful ideas. Basic Books, New York

Partnership for 21st Century Skills (2008) 21st Century Skills, Education & Competitiveness: A Resource and Policy Guide. Tucson: Partnership for 21st Century Skills

Silva E (2008) Measuring skills for the 21st century. Education Sector, pp 1–10. http://www.educationsector.org/publications/measuring-skills-21st-century. Accessed 28 Nov 2010

Smith JP, diSessa AA, Roschelle J (1993) Misconceptions reconceived: a constructivist analysis of knowledge in transition. J Learn Sci 3(2):115–163 (1993–1994)

Todd R (1999) Design and technology yields a new paradigm for elementary schooling. J Technol Stud 25(2):26–33

Vande Zande R (2007) Design education as community outreach and interdisciplinary study. J Learn Through Arts 3(1):1–22

Vygotsky LS (1934/1976) Thought and language. MIT Press, Cambridge, MA

Wallace A (1956) Revitalization movements: some theoretical considerations for their comparative study. Am Anthropol 58:264–281

The Faith-Factor in Design Thinking: Creative Confidence Through Education at the Design Thinking Schools Potsdam and Stanford?

Birgit Jobst, Eva Köppen, Tilmann Lindberg, Josephine Moritz, Holger Rhinow, and Christoph Meinel

Abstract In this chapter, we discuss the idea of "creative confidence" as an objective of design thinking education as taught at the design thinking schools in Potsdam and Stanford. In brief, creative confidence refers to one's own trust in his creative problem solving abilities. Strengthening this trust is a main goal of the education at the design thinking schools. However, there have been only few efforts to develop the concept of creative confidence in design thinking on a deeper and measurable level. To substantiate this discussion, we will compare creative confidence with the concept of self-efficacy and discuss this in the context of the education at the design thinking schools.

1 Introduction

The dominant role of expert knowledge as a resource for professional problem solving has come more and more under pressure since the rise of the information age. The increasing complexity and wickedness of problems we are faced with in our professional lives call for creative and empathic problem solving skills that apply not only a "scientific" knowledge base, but also a generally widespread understanding of various knowledge domains beyond one's own profession. For instance, project-based and multidisciplinary team work as a popular means in the corporate world asks for a greater ability and awareness of sharing and learning

B. Jobst • E. Köppen • J. Moritz • H. Rhinow • C. Meinel (✉)
Hasso-Plattner-Institut (HPI), für Softwaresystemtechnik GmbH, Prof.-Dr.-Helmert-Str. 2-3, Potsdam 14482, Germany
e-mail: Birgit.Jobst@hpi.uni-potsdam.de; Eva.Koeppen@hpi.uni-potsdam.de; Josephine. Moritz@hpi.uni-potsdam.de; Holger.Rhinow@hpi.uni-potsdam.de; meinel@hpi.uni-potsdam.de

T. Lindberg
Hasso Plattner Institute for Software Systems Engineering, Prof.-Dr.-Helmert-Str. 2-3, Potsdam 14482, Germany

H. Plattner et al. (eds.), *Design Thinking Research*, Understanding Innovation,
DOI 10.1007/978-3-642-31991-4_3, © Springer-Verlag Berlin Heidelberg 2012

knowledge from other professions as well as from various stakeholder domains. Skills that help to learn and to transform unfamiliar kinds of knowledge become likewise important for problem solving as skills to apply already internalized knowledge. The question is how companies and employees can attain and maintain these forms of problem solving abilities that lie beyond mere professional training.

Design thinking methodology as taught in at the design thinking schools in Potsdam and Stanford aims at fostering such abilities of meta-professional learning and creativity. "Design Thinkers" are trained in understanding and creatively transforming cross-domain knowledge as well as integrating different expert domains in creative problem solving processes. A core claim of design thinking education is to build up a person's trust in tackling problems of which you rather know what you don't know than what you actually know: This trust in one's own creative capability within a uncertain setting is what we call *creative confidence*. The importance of creative confidence in design thinking has been made clear by David Kelley (2010), founder of the design agency IDEO and one of the "fathers" of design thinking education, by stating that design thinking rather evokes creativity than creating it:

> To me, design thinking is basically a methodology that allows people to have confidence in their creative ability. Normally many people don't think of themselves as creative, or they think that creativity comes from somewhere that they don't know—like an angel appears and tells them the answer or gives them a new idea. So design thinking is hopefully a framework that people can hang their creative confidence on. We give people a step-by-step method on how to more routinely be creative or more routinely innovate. (. . .) And design thinking is basically a method that allows people to have confidence in their creative ability. (Interviewed by Carl von Zastrow 2010)

Also, Rauth et al. (2010) identify through an interview-based study creative confidence as the main learning d.school teachers want to teach. According to the authors, methods, process models and working modes are not seen as a means to foster directly innovative products, but mindsets fostering creative confidence.

However, the questions remain open as to what precisely creative confidence is, that is, how can it be conceptualized, and in which ways design thinking education can reach this goals of fostering it. In this paper, we will address these questions in connection with Albert Banduras concept of self-efficacy. Although the term creative confidence is only vaguely defined, there seems to be a strong similarity with the concept of self-efficacy. Bandura defines self-efficacy as follows:

> Perceived self-efficacy refers to beliefs in one's capabilities to organize and execute the courses of action required to produce given attainments (Bandura 1997).

Self-efficacy therefore supplies the necessary conditions for taking action under risk. If we don't expect success, we will not act or take risks. The same is basically true for creative confidence: If we approach a creative problem without substantiated optimism, it is unlikely that our project will end up being successful. Successful problem solving therefore is not only a result of the amount of knowledge a person has already internalized, but, as Bandura puts it, of *belief*:

Beliefs of personal efficacy constitute the key factor of human agency. If people believe they have no power to produce results, they will not attempt to make things happen (Bandura 1997).

This statement has fundamental implications, meaning that even if we are able to implement a required action we already know about, we will perhaps not do it because we believe that we lack the necessary capacity to succeed. Bandura puts that as follows:

People's beliefs in their efficacy have diverse effects. Such beliefs influence the course of action people choose to pursue, how much effort they put forth in given endeavors, how long they will persevere in the face of obstacles and failures, their resilience to adversity, whether their thought patterns are self-hindering or self-aiding, how much stress and depression they experience in coping with taxing environmental demands, and the level of accomplishments they realize (Bandura 1997).

This clarifies that self-efficacy beliefs influence many motivational, action leading, cognitive and affective processes of a human being by mentally anticipating goal-focused learning processes and estimating their own competence in sufficiently coping with a situation (see also Satow 2002). Self-efficacy therefore can be seen as a crucial precondition for coping successfully with complex challenges in the most diverse fields, regardless of the real individual level of skills.

However, Bandura defines self-efficacy as a general and non-area-specific concept and thus as applicable to diverse situations and indicates therefore that self-efficacy beliefs might vary regarding specific areas. A person, for example, can have a high self-efficacy in an academic context, but the person may have low performance in sports or socializing. This may be the reason why the concept of self-efficacy has been also applied area-specifically, in particular in the field of creativity by, among others, Tierney and Farmer (2002). In this context, the concept of *creative self-efficacy* came up, as stated by Tierney and Farmer (2002):

Working from Bandura's general definition of self-efficacy as targeted perceived capacity, we defined creative self-efficacy as the belief one has in the ability to produce creative outcomes. (Tierney and Farmer 2002)

Against this background, we suggest that the construct of self-efficacy can be used to conceptualize creative confidence in the context of education at the design thinking schools in Potsdam and Stanford. If this assumption turns out to be valid, it would open new perspectives on further research on education at the design thinking schools in Potsdam and Stanford: Bandura's construct of self-efficacy is already operationalized and would allow us to employ validated measurements for the success of education at the design thinking schools. Main similarities between both concepts suggest the soundness of this assumption:

– The base of both terms, creative confidence and self-efficacy, is that people need beliefs in their own capacity before they are able to activate it to the best of their potential.
– Self-efficacy is related to the routine performance of tasks and exercises that generate beliefs through experience. Also education at the design thinking schools aims at generating creative confidence through routine application of

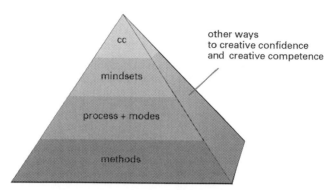

Pyramid Model *by Rauth, Köppen, Jobst, Meinel (2010)*
The development of creative confidence and creative
competence (cc) through design thinking and other
disciplines.

Illustration 1 Pyramid model (Rauth et al. 2010)

methods, process steps and working modes within complex problem settings (see Illustration 1).

However, in spite of these similarities, to further substantiate the connection between self-efficacy and creative confidence in design thinking education, we have to discuss the parallels between both concepts in greater detail. In order to do so, we will depict in the following what sources of self-efficacy Bandura describes and discuss in what way they can be found in design thinking education.

2 Sources of Self-Efficacy

Bandura performed research on how self-efficacy originates and which factors have an impact on self-efficacy. He identifies and describes four sources of self-efficacy. In the following, we will illustrate these sources and then, connecting them to creative confidence, we transfer the sources into the design thinking schools in Potsdam and Stanford by interpreting our explorative observations. Our aim is to check if there are situations and conditions in the design thinking schools in Potsdam and Stanford, which will show that the mediation and enhancement of self-efficacy at the design thinking schools is plausible. The goal of our comparison is therefore to reassess carefully if we find evidence for using the construct of self-efficacy as a synonym to creative confidence in the future.

According to Bandura, the self-efficacy of a person originates from four sources of information: (1) enactive mastery experience, (2) vicarious experience, (3) verbal persuasion and (4) psychological and affective states. In the following, we will compare these four sources with explorative observations at d.schools. Our comparison will adjust each of the four sources separately with at the design thinking schools in Potsdam and Stanford situations in order to find out more about potential parallels and differences.

2.1   Enactive Mastery Experience

Following Bandura, acting out and mastering a difficult task is the first and most efficient way that leads to self-efficacy. Situations that offer direct experiences therefore are a good way to achieve a stronger belief in one's own capabilities. Bandura calls such experiences "mastery experiences" and claims that:

> Successes build a robust belief on one's personal efficacy. Failures undermine it, especially if failures occur before a sense of efficacy is firmly established. (...) After people become convinced that they have what it takes to succeed, they persevere in the face of adversity and quickly rebound from setbacks. By sticking it out through tough times, they emerge from adversity stronger and more able (Bandura 1997).

Also, Bandura points out that even small successes can help people to believe in their own capability to master future tasks or new activities in settings that are uncommon for them (Bandura and Rachman 1978). We therefore asked ourselves: What kinds of difficult situations are posed to students at design thinking schools to be successfully mastered? Which methods are mediated and will empower students to deal successfully with difficult and challenging tasks in the future?

In design thinking education, students get to know the methodological design thinking process by repeating the methods during several so called "design challenges". These design challenges are real projects, which are handed in by project partners. Oftentimes, project partners have different places of origin and come for instance from business or social organisations. But in all cases they hand in a project that has to deal with complex or "wicked" problems (Rittel and Webber 1973). Dealing with ambiguity and wickedness of problems is therefore a main skill that has to be trained during the course of a design thinking education. Finding solutions for wicked problems does not seem to be a promising way for "small" successes that can be easily achieved, as claimed by Bandura. What tools are being delivered by the design thinking education in order to establish creative confidence within students?

We could observe some crucial aspects of methods that may help students to feel more creative and confident. For example, they learn to apply research methods such as interviewing and observation to better know the user and his needs: They learn how to ask and how to observe the user in order to gain empathic knowledge about the user that he himself does not know about or what he cannot elucidate himself. Students then develop a working hypothesis regarding the user's needs, building on the findings and insights of their research. In this phase they develop drawing skills and brainstorming techniques. These ideas from the brainstorming phase are being refined as solution proposals and are made tangible as prototypes. In this way, developing prototyping skills also comes along with the design thinking education.

Throughout the project, students are enabled by several mediated techniques and the knowledge of how to apply these techniques. If we compare our observations with Bandura, we can assume that these techniques help the students to enlarge their problem perspective and to deal confidently with ambiguities within the design challenge. The design thinking methods are tools that are easily achieved and lead to moments of success within the team – the success in problem solving within projects may therefore enable the enhancement of creative self-efficacy.

Bandura also stresses that mastery experience attributes need to be ascribed to one's own capabilities or one's own learning engagement if self-efficacy is to be established. The next question therefore is: What mirrors to the students that the accomplished action was successfully done?

We found that each process step or mode should be shown in a presentation that is open to feedback from the other teams. We could see that the students learn via their presentations and via the given feedback that they are able to solve tasks in a desirable way for the project partner. Consequently, they are more self-confident when it comes to the final presentation in front of the project partners. We could see that the students feel appreciated by these external partners, who are often well-positioned people in an organization. This is an important form of success that comes along with a design challenge. It may even happen that project partners offer concrete jobs to students at the end of a project or that a company financially support a team of alumnis to continue their work on the idea in order to introduce this idea onto the market. Another form of direct success can occur in the form of patents and awards.

Summing up, we assume that design thinking students are in a better position to aquire positive master experiences compared to classical university students. This may be so due to their concrete project experiences including particular familiarity with process models and methods, their teamwork, their specific learning environment and the support of design thinking teachers. According to Bandura, these positive mastery experiences lead to heightened self-efficacy.

2.2   Vicarious Experience

Drawing conclusions about one's own competences are possible when the individual watches other people, for instance models during their acting. The so called "vicarious experience" or "social learning" means that knowledge and cognitive and social skills on the one hand can be acquired by solving problems in teams or on the other hand by watching successful behavioral models, which due to different accounts (insistent effort, effective assignment of learning strategies) can deal with difficult problems and demands. As Bandura describes it:

> The greater the assumed similarity, the more persuasive are the models' successes and failures. If people see the models as very different from themselves, their beliefs of personal efficacy are not much influenced by the models' behaviour and the results it produces. Self-modeling, in which people observe their own successful attainments achieved under

specially arranged conditions that bring out their best, is directly diagnostic of what they are capable of doing.

We assume that students at the design thinking schools are rather similar in their interests (e.g. having an interest in design-oriented approaches). Similar interests also increase identification within teams and therefore enhance social learning as described by Bandura. Apart from having similar interests, the students come from different backgrounds. Due to different study fields, the students expand or obtain different skills, special knowledge, various working methods and other perspectives. A core part of design thinking education is to learn to treat these various knowledge and ability domains complementary and to be open to learning from others. There is therefore very little individual competition in design thinking schools. The attitude of helping each other within and between teams predominates. Teams do not focus on competing with each other but on solving complex challenges and delivering satisfying results.

According to our observations, the diverse teams develop a feeling for the different backgrounds and skills during a project quite well. A psychologist in a multidisciplinary team might bring in his skills to depict mental models and needs of users comprehensively while a product designer might be the only one able to create concept sketches in a fast and comprehensive way. The more they identify themselves with their team members, the better they start to complement one another intuitively. At the design thinking schools in Potsdam and Stanford a team member therefore constantly has experiences that he alone never would have had: experiences of communication, visualization, structuring contents, organisation, risk taking, manifold learning etc. He learns how to observe others – his team members, users, stakeholders – and likewise how it is being observed. Due to the distribution of competences, expertise, skills and ideas among the students, every single one of them moves in a steady flow of vicarious experiences.

Thus teachers are not only instructors of the method but models to the students, as well. They are often involved in design thinking projects and present their results to the students. In comparison to other forms of teaching, the design thinking schools in Potsdam and Stanford are characterized by an open atmosphere also concerning the relationships between teachers and students. Since teachers are not judging or evaluating the students, they can act as advisors, models, sometimes just cocreators that give useful hints. According to Amabile (1996b) the creativity is enhanced additionally if one works together with a "coactor" that reflects the team's or individuals creative outcomes. Teachers as "coactors" therefore serve as a source for vicarious experiences as well. Also the use of open spaces and flexible working and communication surroundings like mobile furnitures and communication supports this process of constantly observing others as models in action in order to reflect their own actions.

To sum it up, vicarious experiences are made in at the design thinking schools in Potsdam and Stanford in various ways. The students learn complementary skills, working methods and behaviors by watching their fellow students and teachers. The d.schools have well functioning and flexible premises, which create free space to bring forward cooperative communication and therefore supports social learning.

A particular culture is promoted which implies small teams with teachers monitoring and supporting the students throughout the entire processes and providing feedback, without judging them too early. This particular atmosphere can be regarded as a class climate which enforces learning. We are convinced that it will affect self-effective expectations in a positive way.

2.3   Verbal Persuasions

A further important source for the development of self-efficacy expectations (auto-suggestion: "You can make it!") refers to verbal feedback or verbal persuasion. Verbal persuasion means that one persuades someone of being capable of doing something in a successful manner. Verbal feedback provided by another party is especially helpful and effective if it occurs task-related and promptly and if it shows realistic consideration of the actual level of skills, abilities and the performed learning progress of the team members (see Kutner 1995; Schwarzer and Jerusalem 2002). Not only the verbal persuasion by other people is effective, but also one by one's own inner voice. The so-called "self-instruction" belongs to this source of self-efficacy as well. As Meichenbaum (1985) describes it, self-instruction and self-verbalization are two of the prominent methods in psychology and specifically in behavioral therapy. They have proven to be valid concepts for handling stressful or frightening situations.The emphasis on self-instruction is placed on the meaning of the conviction of the own capacity for acting ("I can do it!", "I have the right to do so!") and is therefore related to the encouragement of self-efficacy. As puts it:

> People who are persuaded verbally that they possess the capabilities to master given tasks are likely to mobilize greater effort and sustain it than if they harbor self-doubts and dwell on personal deficiencies when difficulties arise. Bandura (1997)

There is a high degree of mutual support and motivation in design thinking teams. Through the use of motivation techniques, an atmosphere of constructive feedback and an attitude towards failure as a means for learning, there is generally a low level of fear and a high level of optimism. For instance, "fail early and fail often" is one of the key paradigms in design thinking and requested and welcome as a chance for further learning. Within the process and the course of the project there exists informal and encouraging feedback at all times. Speaking out of experience, a strong belief in the capabilities of a design thinking team generally goes along with an attitude of "Yes, we can do it!"

Also, the design thinking school environment offers strong social support, in particular through the teachers. At least one teacher is assigned to every design thinking team. He or she mentors and accompanies the students throughout the whole process and during the entire period of the project. If required, the teacher passes the team through certain project phases and intervenes if the process stagnates or if methods are applied incorrectly or in an unhelpful way. If a team does not get along well within the process and makes no progress or is not capable of changing this status, the teacher joins in and supports the team in passing the actual process phase by asking the aim oriented questions, reflecting on the situation and by giving the team further methods to continue with. Moreover, some teachers actively participate in presentations (e.g. they take over a role within the role play). Other teachers also take part in activities outside the regular lessons. In this way, the teachers act as guides providing the team with a feeling of backup throughout the process.

2.4 Physiological and Affective States

Physiological and affective states as well as physical arousal are expressions of the perceived belief in one's own self-efficacy and influence one's expectancy of self-efficacy. Self-efficacy is also influenced by the observation of one's own emotional states while a person thinks about a certain task or tries to solve a problem.

> People often read their physiological activation in stressful or taxing situations as signs of vulnerability to dysfunction. Because high arousal can debilitate performance, people are more inclined to expect success when they are not beset by aversive arousal than if they are tense and viscerally agitated. Stress reactions to inefficacious control generate further stress through anticipatory self-arousal. By conjuring up aversive thoughts about their ineptitude and stress reactions, people can rouse themselves to elevated levels of distress that produce the very dysfunctional they fear.

In the design thinking schools in Potsdam and Stanford, every day starts with so-called "warm-ups" to relax the team members. Since Jacobson (1938) we know that activity in the central nervous system influences the muscular tension and vice versa. That means, psychological strain goes along with an increased tonicity, but, on the other hand, warm-ups also lead to the relaxation of the musculature and contribute to mental stress relaxation (Esser Göggerle and 2008).

Even through the warm-ups in the design thinking schools do not seem to be specifically focused on relaxation training and mental relaxation, one can assume that these practices contribute to a decrease of nervousness and stress and negative or disruptive thoughts and feelings take a back seat.

The following tasks and challenges may be accomplished more easily with this state of mind. A heightened dumping of endorphin additionally leads to individual satisfaction and enhancement of motivation.

Not only are relaxation and the reduction of pressure important consequences of physical exercises – at the same time the common acting within the group seems to heighten a certain "we-feeling" and team spirit.

In addition, warm-ups are created in a way that small tasks have to be performed. Because of the low complexity factor of these tasks (e.g. "create a new greeting procedure and greet your neighbor with it") the participants gain a feeling of success right from the beginning. This provides relaxation but is also a convenient contrast to the many small failures the teams will have to face in their projects.

We can summarize that the fourth source of self-efficacy can only be conveyed indirectly by the design thinking education. Nevertheless, the design thinking schools uses warm-ups that can lead to a decrease of stress reactions. Within a comfortable atmosphere, with moments of success and social support from the other group members, negative affective states will occur more infrequently. We therefore assume that this fourth source of self-efficacy is nevertheless being addressed by the design thinking schools and that it has got a positive influence on the self-efficacy of the students.

3   Conclusion

In this paper, we explored whether the concept of creative confidence in the education at the design thinking schools in Potsdam and Stanford can be conceptualized by Bandura's concept of self-efficacy. We thereby compared Bandura's concept of four sources of self-efficacy with key aspects of the design thinking schools. It has been shown that compared to rather traditional forms of learning and education it is more likely that students achieve positive mastery experiences in an easy way. Also students at the design thinking schools are supported by social learning, constant feedback by the teachers and other students as well as a constructive atmosphere. The focus on physical exercises such as warm-ups in the d.school context that are performed by the team, also led to the conclusion that stress reactions and fearful moments are decreased on a physiological base.

In summary, all four sources of self-efficacy proposed by Bandura can be found in core aspects of d.school education. We therefore conclude that the concept of creative confidence as a main goal to be taught at d.schools can be – at least to a high degree – conceptualized through the self-efficacy construct. Both creative confidence and self-efficacy refer to one's own trust in his creative problem solving abilities and are built upon the idea that successful experiences will have positive effects on future challenges. We acknowledge the specific context of creative confidence in design thinking schools. Therefore we assume that Tierney and Farmer's (2002) definition of creative self-efficacy can be regarded as a promising starting point to further explore the impact of design thinking on the teams' attitudes and behaviours. However, this does not mean that Bandura's concept does completely define creative confidence in the design thinking schools. There are probably more factors to consider like personal educational backgrounds or task-related issues. Nevertheless, the parallels between both concepts are salient and future research on the effectiveness of design thinking schools should take this into consideration.

4   Outlook

Creative confidence is an objective of the design thinking schools in Potsdam and Stanford and an important skill for future generations of students. Design thinking education intends to mediate this capability next to other crucial skills such as "wicked" problem solving and empathic learning abilities. In order to investigate the actual effects of design thinking education, it seems to be promising to use the construct of self-efficacy as a means of measuring the amount of creative confidence students gain through design thinking schools courses. Our goal therefore is to design a research framework drawing upon the already validated measurements of self-efficacy.

We have already completed some pilot studies and gained some first quantitative and qualitative data. The insights from the data and the current comparison of Bandura's sources at the design thinking schools in Potsdam and Stanford support our assumption that design thinking schools in Potsdam and Stanford conveyes creative self-efficacy. We presume that the effects and factors of self-efficacy are the same as for creative self-efficacy. In the framework of a larger research project, we will evaluate empirically if design thinking schools mediate creative self-efficacy and if so, what influential factors can be observed. We hope to better understand the mediation of this skill to be able to give suggestions for the design thinking education as well as for the development of creative confidence in other contexts.

References

Amabile TM (1996b) Creativity in context. The social psychology of creativity. Boulder, Colo.; Oxford: Westview Press

Bandura A, Rachman S, (1978) Perceived self-efficacy: analyses of Bandura's theory of behavioural change. Oxford, Eng.; New York, N.Y.: Pergamon Press

Bandura A (1997) Exercise of control. W. H. Freeman and Company, New York

Esser G, Göggerle S (2008) Lehrbuch der klinischen Psychologie und Psychotherapie bei Kindern und Jugendlichen. Georg Thieme Verlag

Jacobson E (1938) Progressive relaxation; a physiological and clinical investigation of muscular states and their significance in psychology and medical practice. Chicago: Univ. of Chicago Press

Kutner L (1995) Die Bedeutung der Selbstwirksamkeit für die Anpassung Jugendlicher an den gesellschaftlichen Wandel. In: Edelstein W (eds) (Hrsg.) Entwicklungskrisen kompetent meistern. Der Beitrag der Selbstwirksamkeitstheorie von Albert Bandura zum pädagogischen Handeln, Heidelberg, pp 74–84

Meichenbaum D (1985) Teaching thinking: a cognitive-behavioural perspective. In: von Segal JW, Chipman SF, Glaser R (eds) Thinking and learning skills: research and open questions, vol 2. Lawrence Erlbaum Associates Inc., Hillsdale, pp 407–422

Rauth I, Köppen E, Jobst B, Meinel C (2010) An educational model towards creative confidence. In: 1st Proceedings of ICDC, Kobe

Rittel HWJ, Webber M (1973) Dilemmas in a general theory of planning. Policy Sci 4(2):155–169

Satow L (2002) Unterrichtsklima und Selbstwirksamkeitsentwicklung. In: Jerusalem M, Hopf D (eds) Selbstwirksamkeit und Motivationsprozesse in Bildungsinstitutionen. Beltz, Weinheim (Zeitschrift für Pädagogik, Beiheft; 44), pp 174–191

Schwarzer R, Jersusalem M (2002) Das Konzept der Selbstwirksamkeit. In: Jerusalem M, Hopf D (eds) (Hrsg.) Selbstwirksamkeit und Motivationsprozesse in Bildungsinstitutionen. Weinheim, pp 28–53

Tierney P, Farmer SM (2002) Creative self-efficacy: its potential antecedents and relationship to creative performance. Acad Manage J 45(6):1137–1148

von Zastrow C (2010) New designs for learning: a conversation with IDEO Founder David Kelley. http://www.learningfirst.org/visionaries/DavidKelley. Accessed 5 Nov 2011

Prototyping Dynamics: Sharing Multiple Designs Improves Exploration, Group Rapport, and Results

Steven P. Dow, Julie Fortuna, Dan Schwartz, Beth Altringer, Daniel L. Schwartz, and Scott R. Klemmer

Abstract Prototypes ground group communication and facilitate decision making. However, overly investing in a single design idea can lead to fixation and impede the collaborative process. Does sharing multiple designs improve collaboration? In a study, participants created advertisements individually and then met with a partner. In the *Share Multiple* condition, participants designed and shared three ads. In the *Share Best* condition, participants designed three ads and selected one to share. In the *Share One* condition, participants designed and shared one ad. Sharing multiple designs improved outcome, exploration, sharing, and group rapport. These participants integrated more of their partner's ideas into their own subsequent designs, explored a more divergent set of ideas, and provided more productive critiques of their partner's designs. Furthermore, their ads were rated more highly and garnered a higher click-through rate when hosted online.

1 Introduction

Many designers live by the principle, "never go to a client meeting without a prototype" (Thomke and Nimgade 2000). Prototypes help people summarize their ideas, demonstrate progress and expertise, surface implicit design vocabulary, and ground group communication and decision making (Buxton 2007; Schon 1995; Schrage 1999). Creating a prototype—sketching a possible future—helps people

S.P. Dow (✉) • J. Fortuna • D. Schwartz • D.L. Schwartz • S.R. Klemmer
Human-Computer Interaction Group, Stanford University, Gates Computer Science Building, 353 Serra Mall, Stanford, CA 94305, USA
e-mail: spdow@stanford.edu; jfortuna@stanford.edu; dschwartz13@stanford.edu; danls@stanford.edu; srk@stanford.edu

B. Altringer (✉)
Harvard School of Engineering and Applied Sciences, Cambridge, MA, USA
e-mail: bethalt@seas.harvard.edu

H. Plattner et al. (eds.), *Design Thinking Research*, Understanding Innovation, DOI 10.1007/978-3-642-31991-4_4, © Springer-Verlag Berlin Heidelberg 2012

see the entailments and interactions of their design ideas and communicates them to other stakeholders (Dow et al. 2009; Gerber 2010).

Rapid iteration provides value, but it does not guarantee broad exploration (Dow et al. 2010). People systematically overestimate the predictability of the future, especially in complex situations (Larrick 2009). For example, when financial experts estimate the range of possible futures, they consistently underestimate the variance (Ben-David et al. 2007). In prediction and decision-making tasks, people can improve the quality of their estimates by broadening the frame and generating multiple guesses under different assumptions (Herzog and Hertwig 2009; Larrick 2009).

We hypothesize that creating and sharing *multiple* alternatives has more benefits than people may realize. Both cognitive and social factors motivate this hypothesis. First, the presence of a concrete prototype may (for better and for worse) focus the discussion on refining *that* idea rather than thinking more broadly (Cross 2004; Jansson and Smith 1991). Without exploration, people often interpret the frame of the design problem too narrowly (Kershaw and Ohlsson 2004). Second, people presenting designs often believe their status to be on the line (Buxton 2007). This risk encourages overinvesting time, labor, psychological energy, and social momentum into a single concept (Brereton et al. 1996; Dow et al. 2010). In this single-prototype strategy, individuals may seek validation for their ideas and disregard or fear the critique and feedback necessary for exploration and revision (Nickerson 1998). Compounding this, collaborative work is often susceptible to *groupthink*, where members reinforce each other's belief in the current direction at the expense of other options (Janis 1982).

Creating multiple prototypes in parallel can help individuals more effectively understand underlying design principles, enumerate more diverse solutions, and react less negatively to feedback (Dow et al. 2010; Nielsen and Faber 1996). Distributing one's psychological investment across multiple designs can reduce fixation and sunk-cost reasoning (Arkes and Blumer 1985; Jansson and Smith 1991). Individuals may be more candid and critical of their own and others' ideas (Davidoff et al. 2007; Tohidi et al. 2006), resulting in more fluid and effective collaboration.

However, creating multiple alternatives leaves less time to polish each single one and may be perceived as wasting effort (Schrage 1999). Focusing on fewer endeavors can help people focus, contemplate, relax, and be more productive (Iyengar and Lepper 2000; Mark et al. 2005). Increasing options can cause analysis paralysis—a "paradox" of choice (Schwartz 2004)—and may jeopardize a group's ability to achieve consensus (Ball and Ormerod 1995).

This paper investigates whether sharing multiple prototypes increases design performance, improves group interaction, and leads to more effective idea sharing. In a between-subjects experiment, 84 participants working in pairs designed Web banner advertisements for a non-profit organization. The study comprised three steps. First, participants prototyped designs individually. Second, they worked with a partner to critique each other's ideas. Third, each individual created a final ad. Participants answered survey questions at several points and open-ended questions

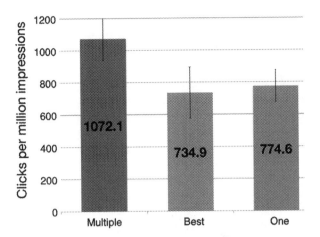

Fig. 1 Online ad performance (clicks per million impressions): *Share Multiple* ads outperformed the other conditions

at the end. Pairs were randomly assigned to one of three conditions: creating and sharing multiple ads; creating multiple ads and sharing the best; and creating and sharing one ad. Comparing these three conditions separates the effects of *producing* multiple designs and *sharing* multiple designs. Each condition was allotted the same time for design.

Ads in the *Share Multiple* condition generated significantly more clicks per impression than the other conditions (see Fig. 1). Independent (and blind-to-condition) judges rated ads from the *Share Multiple* condition significantly higher. Judges also rated *Share Multiple* ads as significantly more divergent. Participants in the *Share Multiple* condition shared significantly more ideas and moved more towards consensus than pairs who shared only one design. Group members in the *Share Multiple* condition reported a greater increase in rapport over the course of the experiment, while rapport in the other two conditions dropped. Moreover, *Share Multiple* participants exchanged speaking turns significantly more often.

In short, sharing multiple designs improves outcome, exploration, sharing, and group rapport. These results suggest that encouraging group members to share multiple ideas will pay dividends in both design outcomes and interpersonal dynamics. The following subsections elaborate the study's rationale and hypotheses.

2 Exposure to Examples Enhances Individual Exploration

Exposing people to examples increases the likelihood they will integrate similar features into their own designs (Smith 1993), even when they are asked to create vastly different ideas (Finke et al. 1996). Furthermore, borrowing increases with the number of examples people see (Marsh et al. 1996). Smith et al. hypothesized that people often take the path of least resistance, and that this conformity constrains creativity (Smith 1993). However, using Smith et al.'s task, Marsh et al. found that participants who saw many examples created equally novel work (Marsh et al. 1996). In other words, participants borrowed from examples when they lacked a better idea, but viewing examples did not "push out" or inhibit people's novel ideas. Furthermore, when viewed from a *quality* perspective, people perform better when examples are readily available (Brandt et al. 2010; Lee et al. 2010). In all of this prior work, examples were presented anonymously. Collaborative work is importantly different in this regard because the examples are produced by a known and co-present peer.

This paper hypothesizes that producing multiple designs and being exposed to multiple examples produced by other group members leads individuals to create a more divergent set of concepts.

Hypothesis 1

Creating and viewing multiple designs leads to more individual exploration.

This study measured individual design exploration by having independent raters judge the diversity/similarity of each participant's designs.

2.1 Sharing Multiple Designs Improves Collaboration

Designers often work collaboratively to generate, critique, and revise ideas, and to build consensus (Gerber 2010; Schon 1995; Warr and O'Neill 2005). Under controlled conditions, individuals working separately often collectively produce a greater volume of ideas than group brainstorming (Diehl and Stroebe 1987; Taylor et al. 1958). Group members may block each other from sharing ideas (Stroebe and Diehl 1994), get frustrated with bad apples in the group (Felps et al. 2006), and "free ride" by deferentially following others' ideas (Janis 1982). However, measuring only the sheer volume of ideas is misleading: group brainstorming supports organizational memory of design solutions, recognizes skill variety among team members, and builds shared ownership of ideas—crucial for selecting and refining concepts (Sutton and Hargadon 1996). To some extent, the debate over whether to design individually or collectively presents a false choice; creative work typically involves both (Bao et al. 2010; Sutton and Hargadon 1996).

Sharing ideas with a group can be an anxiety-laden experience, and this anxiety can negatively affect performance (Dannels and Martin 2008). Individuals who know they will be judged by experts produce less novel ideas (Diehl and Stroebe 1987). Many critique providers are aware that public feedback can be emotionally fraught; consequently they take care to temper criticism (Tohidi et al. 2006) and supplement critique with praise (Hyland and Hyland 2001). Anxiety may increase when people believe their worth as a person is part of what's being assessed (Dannels and Martin 2008; Kosara 2007). For this reason, many educators and parents use language that critiques the *work* and the *behavior*, rather than the *person* (Dweck 2007).

Creating multiple designs may help both critiquers and creators separate egos from artifacts. When asked for feedback, people provide more substantive critique when presented with multiple design alternatives (Tohidi et al. 2006). People react less negatively when they receive critique on multiple alternatives in parallel (Dow et al. 2010). This prior work studied individual behavior; this paper analyzes the social effects.

This paper hypothesizes that sharing multiple designs—rather than one—improves group rapport and increases the rate at which people exchange ideas.

Hypothesis 2

Sharing multiple designs leads to more productive dialogue and better group rapport.

This study measured peer interaction by counting speech turns by each partner (Ranganath et al. 2009). Also, five questions posed before and after the group discussion assessed individual views of their group's rapport.

2.2 *Sharing Multiple Ideas Facilitates Conceptual Blending*

When collaborating, groups often merge properties of different concepts (Fauconnier and Turner 2003). Sometimes, these blends directly inherit properties (Hampton 1987), other times blends spawn new emergent features (Finke et al. 1996; Schwartz 1995). Blending can be highly structured, as in morphological design (Zwicky 1969), but is more commonly ad hoc. When concepts are dissimilar, blending them yields a more ambiguous artifact (Wisniewski and Gentner 1991).

Conceptual ambiguity can beneficially provide a generative resource (Gaver et al. 2003; Leifer 2010). Sharing multiple designs may help collaborators blend ideas. The process of comparing and contrasting alternatives helps people create higher-level structures (Thompson et al. 2000); these structures help collaborators understand and communicate the rationale behind design decisions (Moran and Carroll 1996).

This paper hypothesizes that sharing multiple design concepts facilitates conceptual blending and that collaborators will use more surface-level and thematic features from their partner's work.

Hypothesis 3

Sharing multiple designs leads to more effective conceptual blending.

This study measures conceptual blending by counting features that migrate from one partner's preliminary designs to the other's final design. Independent raters also judged the similarity of partner's designs before and after the pair shared their work.

Finally, this paper hypothesizes that sharing multiple designs leads to better performance due to a confluence of three factors: individuals explore more divergent ideas; groups have stronger dialogue and rapport; and the final design exhibits more effective conceptual blending.

Hypothesis 4

Sharing multiple designs produces better results.

This study measures design quality by gathering click-through performance metrics on advertisement designs and by recruiting professionals, clients, and other independent judges to rate ads.

3   Method

A between-subjects study manipulated the prototyping process prior to a group critique. Web advertising was chosen as the design task because it fulfills key criteria:

- Quality can be measured objectively and subjectively;
- Participants need minimal artistic or engineering ability to either create or critique ads;
- Individuals can complete tasks within a single lab session;
- Solutions demonstrate creative diversity and a range of performance quality.

3.1   Study Design

Participants all created Web ads for the same client, *FaceAIDS.org*. The study allocated equal time for individual design and group discussions across three conditions. In the *Share Multiple* condition, participants created three preliminary advertisements and shared all three during the group discussion. In the *Share Best* condition, participants created three preliminary ads and chose one to share during the group discussion. In the *Share One* condition, participants spent the entire individual design time on a single ad to share during the group discussion.

3.2   Participants

We recruited 84 participants by way of fliers, online advertisements, and email lists. Two individuals arrived concurrently to form study pairs. Each pair was assigned to one of three conditions using a stratified randomization approach; the study balanced for gender (41 females) and graphic design knowledge across pairs and conditions. Ten true-or-false questions assessed graphic design knowledge (see Appendix A); participants were deemed *experienced* if they correctly answered eight or more (36 did). Participants who scored below eight were deemed *novices*. Participants' average age was 26.5; three-fourths were students.

3.3   Procedure

The experiment comprised the following steps: consent form, icebreaker, tool training, practice ad, design brief, individual design, group discussion, final individual design, group interview, and final debriefing. Questionnaires collected demographic and self-report assessments. The icebreaker, group discussion, and group interview were co-located and video-recorded. All other procedures took place in separate rooms at individual workstations with no video recording. For 120 min of participation, subjects received $20 USD cash.

3.4   Icebreaker Activities

Partners collaborated on three icebreaker activities for 3 min each. They built a tower with toy blocks, played the game Operation, and generated a list of animal names beginning with 'M' (e.g., monkey).

3.5   Graphic Design Tool Training

At separate workstations, partners viewed a 5-min video about the Web-based graphic design tool (http://flashimageeditor.com). Then, using the tool, participants replicated a graphic unrelated to the main task. All participants replicated the graphic in less than 10 min. None had used the tool before. Selecting a novel tool avoids confounds from participant's tool-specific expertise.

3.6   Design Brief

A 5-min video described participants' main design activity: to create an ad for *FaceAIDS*, a non-profit organization dedicated to global health equity and social justice (http://faceaids.org). In the video, the organization's executive director outlined four goals: reach out to students interested in starting local chapters of *FaceAIDS*, increase traffic to the *FaceAIDS* Web site, impress three judges from the *FaceAIDS* organization, and create ads with effective graphic design. A paper version of the design brief was available for the group discussion (see Appendix A).

3.7   Individual Design Period

All participants had 30 min for individual design. In the *Share Multiple* and *Share Best* conditions, participants started a fresh design every 10 min. This was typically adequate, even for novices. At the end of this period, *Share Best* participants were prompted to select one design to be critiqued by the study partner. After the design period, a study proctor printed ads for the group discussion.

3.8 Group Discussion

Participants sat together and viewed a printout of their partner's design(s). The study proctor set a timer for 5 min and then instructed the pair: "Examine your peer's design concept(s) and then provide a critique. What advice would you provide? Please speak aloud." After this, the proctor set another 5-min timer and instructed: "Now spend another 5 min discussing what you think is the most effective way to satisfy the design brief." After that, participants were instructed to return to their individual workstations to create a final ad design.

3.9 Final Design Period

Participants individually created another advertisement and were instructed that this final ad would be rated by judges and hosted in a live ad campaign.

3.10 Group Interview

The study concluded with an open-ended group interview. A study proctor used an interview guide and followed up with related questions. These questions provided guidance for the final interview; the exact order and phrasing varied.

- Describe how you arrived at your final design.
- Explain the difference between your two final ads.
- How much did the group discussion affect what you did in your final ad design?
- How did your peer's critique affect your ad design?
- To what extent were you able to reach agreement on the final design concept?

4 Dependent Measures

4.1 Performance

After the experiment, the final graphic ads were hosted on Google AdWords (http://adwords.google.com) for a 12-day campaign. Design performance was determined through two objective measures:

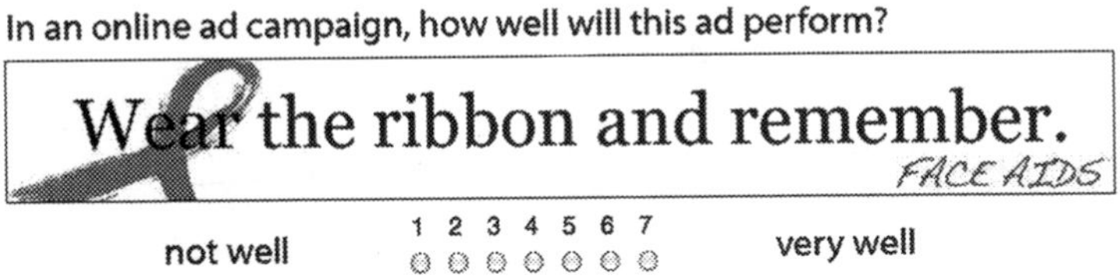

In an online ad campaign, how well will this ad perform?

Fig. 2 Quality rating: judges rated (on a 1–7 scale) how well each ad will perform in an online campaign

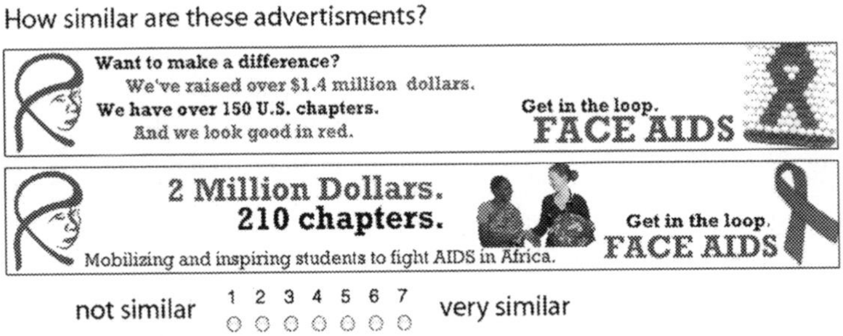

How similar are these advertisments?

Fig. 3 Similarity rating: judges viewed a pair of ads and rated their similarity on a seven-point scale. This pair's average similarity rating was 5.7. (The overall average was 3.6)

- Click-through rates (CTR): number of clicks divided by the number of impressions, and
- Google Analytics (http://www.google.com/analytics) on the target client Website: total time spent and number of pages visited from each ad.

Ads were also independently judged by 30 individuals: 3 clients from *FaceAIDS*, 6 ad professionals, and 21 people recruited from Mechanical Turk, an online crowdsourcing system for paying workers for short tasks (http://mturk.com/mturk/welcome). This collection of raters provided important—and different—audience perspectives. Each judge read the *FaceAIDS* design brief and viewed ads in random order. For each ad, they estimated (on a seven-point scale) each ad's performance in an online campaign (see Fig. 2).

4.2 Individual Design Exploration

Exploring a diverse set of ideas can help people examine the space of designs and their relative merits (Buxton 2007). To obtain a measure of idea diversity, ten independent raters assessed pair-wise similarity of all combinations of individual participant's ads (see Fig. 3). Raters recruited from Amazon Mechanical Turk assessed similarity on a scale from 1 to 7 (*not similar* to *very similar*).

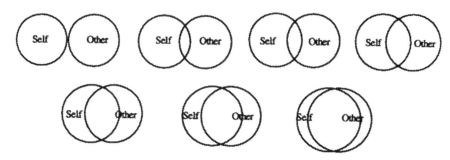

Fig. 4 Inclusion of self in others scale: illustration reprinted from Aron et al. (Brandt et al. 2010)

4.3   Change in Group Rapport

At two points—after the icebreakers and after the discussion—five questions asked individuals to assess their group rapport. The change between these two points measures the discussion's impact on group rapport. Four questions originate from the Subjective Value Inventory (SVI), an assessment of viewpoints on negotiation (Curhan et al. 2006). The relationship questions from the SVI provide a systematic measure of a group rapport; they assess partners' feelings about the relationship in terms of overall impressions, satisfaction, trust, and foundations for future interaction. The fifth question derives from the Inclusion of Self in Others Scale (see Fig. 4), a measure of someone's sense of connectedness with another (Aron et al. 1992). The questions asked:

- What kind of overall impression did your peer make on you?
- How satisfied are you with your relationship with your peer as a result of the interaction?
- Did the interaction make you trust your peer?
- Did the interaction build a foundation for future interactions with your peer?
- Please check the picture below which best describes your relationship with your peer:

4.4   Conversational Turn Taking

In the group discussion, partners exchanged ideas. A coder recorded the start time and duration of each group member's utterances. This provided the overall number of speech turns by each partner, the total amount of speaking, the ratio of time spent by the more conversationally-dominant partner (high and low talkers), and the frequency of turns per minute of interaction.

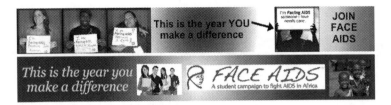

Fig. 5 Design feature sharing: The two partner ads above have **three** commonalities: images, phrasing, and background color

4.5 Design Feature Sharing

For each final ad, we counted cross-pollinated features in five categories: word phrases, background color, images, layout, and styles (i.e., fonts, rotations, etc.). Cross-pollination was a binary value for each category. A category received a mark if a participant's final ad exhibited a feature that was present in their partner's shared provisional ad, but *not* in their own provisional ad(s) (see Fig. 5).

4.6 Group Consensus

As an aggregate measure of group consensus, independent raters assessed pair-wise similarity between partner ads. The similarity assessment contrasted ads created before and after the discussion. Ten raters recruited from Mechanical Turk assessed similarity on a scale from 1 to 7 (not similar to very similar) (see Fig. 3). If the designs are more similar after the discussion, it suggests that partners converge around similar concepts.

5 Results

Participants created a wide variety of ad designs, demonstrating a range of quality. The highest-rated ads tended to be original, visually appealing, and cleverly touched on themes relevant to *FaceAIDS* (see Fig. 6, left column). Ads with high click-through rates grabbed the attention of web users by employing more unconventional color palettes, layouts, and rhetorical hooks (see Fig. 6, right column). Highly rated ads did not correlate with high performing ads ($R^2 = 0.018$, $F(1,79) = 2.445$, $p > 0.05$, $b = 0.174$). The third highest performing ad, for example, was the second lowest rated ad. This paper considers different meanings of quality by examining various outcomes.

FaceAIDS reviewed the ads before they appeared online. The client found four of the ads to have inappropriate negative imagery, and requested they not be shown. Three of these were from *Share Best;* one was from *Share Multiple*. In total, the ad campaign generated 239 clicks on 274,539 impressions (ad appearances). The total advertising costs were $362 USD (an average of $1.51 per click).

Fig. 6 Top five highest-rated ads (*left*); ads with top five click-through rates (*right*)

The results supported all four hypotheses. Participants in the *Share Multiple* condition produced higher-quality designs (better click-through rates and higher ratings) and created more diverse designs. Pairs in the *Share Multiple* condition reported a greater increase in rapport, exchanged more verbal information, and shared more ideas. Moreover, ads by more experienced participants received higher ratings than novices; the designs created by experienced participants were less diverse than novices.

6   Sharing Multiple Led to Higher Quality Designs

6.1   Ad Campaign Results

A chi-squared analysis examined ad campaign performance for all 12 days. *Share Multiple* ads had 98,867 impressions with 106 clicks, *Share Best* ads had 77,558 impressions with 57 clicks, and *Share One* ads had 98,038 impressions with 76 clicks (Fig. 1 summarizes the average clicks per million impressions). *Share Multiple* ads had a significantly higher click-through rate ($\chi^2 = 4.72$, p < 0.05).

An analysis of variances was performed with condition (*Share Multiple*, *Share Best*, and *Share One*) and graphic design score (*experienced* or *novice*) as factors and total time spent and pages visited as dependent variables. Between conditions, there were no differences for total time spent (F(5,202) = 0.808, p > 0.05) or number of pages visited from each ad (F(5,202) = 0.461, p > 0.05).

6.2   Graphic Design Experience Did Not Effect Campaign Results

Ads created by participants who scored high on the graphic design exam garnered 110 clicks on 128,783 impressions; novice ads had 129 clicks on 145,756 impressions. This was not a significant difference ($\chi^2 = 0.08$, p > 0.05). Experienced participants benefited more from the manipulation than novices did

Table 1 Online performance in clicks per million impressions for condition and experience (std dev in parentheses). Experienced participants created better ads and were more affected by condition

	Experienced	Novice
Share Multiple	1,125.3 (818.0)	991.7 (482.2)
Share Best	704.3 (422.2)	758.9 (888.5)
Share One	540.3 (500.0)	905.5 (371.7)

Table 2 Average ratings (std dev in parentheses). All rater types (clients, ad pros, and turkers) rated *Share Multiple* ads higher

	Clients	Ad pros	Turkers
Share Multiple	4.06 (1.70)	2.95 (1.63)	4.14 (1.80)
Share Best	3.45 (1.77)	2.76 (1.63)	3.90 (1.74)
Share One	3.79 (1.69)	2.85 (1.55)	3.94 (1.68)

(see Table 1). Experienced participants in the *Share Multiple* condition outperformed experienced participants in the *Share Best* ($\chi^2 = 3.95$, $p < 0.05$) and *Share One* conditions ($\chi^2 = 8.33$, $p < 0.05$). There were no ad performance differences between conditions for novices.

An ANOVA showed that experience did not significantly affect total time spent ($F(5,202) = 0.091$, $p > 0.05$) or number of pages visited from each ad ($F(5,202) = 0.076$, $p > 0.05$).

6.3 Quality Ratings

Thirty raters judged all final ads on a seven-point scale (1 = poor and 7 = excellent). An analysis of variances was performed with condition (*Share Multiple, Share Best,* and *Share One*) and graphic design score (*experienced* or *novice*) as factors and performance rating as the dependent variable. The *Share Multiple* condition ($\mu = 3.89$, $SD = 1.82$) outperformed the other conditions ($F(2,2519) = 5.075$, $p < 0.05$). The difference between the *Share Best* ($\mu = 3.63$, $SD = 1.78$) and *Share One* ($\mu = 3.71$, $SD = 1.71$) conditions was not significant ($p > 0.05$; Tukey's test).

Professionals, clients, and turkers (workers on Amazon Mechanical Turk) all rated *Share Multiple* ads higher than the other conditions (see Table 2). Clients ($\mu = 3.77$, $SD = 1.73$) and turkers ($\mu = 3.99$, $SD = 1.74$) ratings were higher on average than those by ad professionals ($\mu = 2.85$, $SD = 1.60$) ($F(2,2519) = 86.961$, $p < 0.05$). This differential between advertising professionals and other stakeholders is consistent with prior work (Dow et al. 2010).

Table 3 Average ratings by condition and experience (std dev in parentheses). Experienced created higher-rated ads; novices were more affected by condition

	Experienced	Novice
Share Multiple	4.11 (1.71)	3.68 (1.90)
Share Best	4.19 (1.71)	3.31 (1.74)
Share One	4.01 (1.71)	3.48 (1.67)

Table 4 Individual views of group rapport rose in the *Share Multiple* condition; it dropped in other conditions (std dev in parentheses)

	Before design critique	After design critique	Group rapport shift
Share Multiple	$\mu = 24.6$ (4.35)	$\mu = 25.5$ (4.66)	+0.89
Share Best	$\mu = 24.0$ (5.24)	$\mu = 22.3$ (4.79)	−1.75
Share One	$\mu = 24.9$ (5.18)	$\mu = 22.8$ (5.86)	−2.11

6.4 Graphic Design Experience Led to Better Ratings

Participants who scored highly on the graphic design exam ($\mu = 4.10$, $SD = 1.709$) significantly outperformed those who scored poorly ($\mu = 3.48$, $SD = 1.773$), ($F(1,2519) = 74.613$, $p < 0.05$). The ANOVA shows that novices benefited more from the manipulation than experienced participants did ($F(2,2519) = 3.536$, $p < 0.05$) (see Table 3). This differential gain is the opposite from the click-through rate, where experienced participants benefited more from sharing multiple designs.

7 Sharing Multiple Led to More Individual Exploration

Raters from Amazon Mechanical Turk deemed *Share Multiple* ads to be most divergent. An analysis of variances was performed with condition (*Share Multiple, Share Best,* and *Share One*) and graphic design score (*experienced* or *novice*) as factors and pair-wise similarity rating as the dependent variable. The similarity rating differed significantly across conditions ($F(2,3640) = 82.07$, $p < 0.05$). Tukey post-hoc comparisons of the three conditions indicate that *Share Multiple* ads ($\mu = 3.85$, $SD = 1.93$) were more diverse than *Share Best* ads ($\mu = 3.99$, $SD = 1.96$) ($p < 0.05$) and *Share Best* ads were more diverse than *Share One* ads ($\mu = 5.45$, $SD = 1.86$) ($p < 0.05$).

Experienced participants created ads that were deemed significantly more similar ($\mu = 4.20$, $SD = 1.96$) than those who scored poorly ($\mu = 3.91$, $SD = 1.99$) ($F(1,3640) = 7.692$, $p < 0.05$). There was no interaction effect between condition and prior experience.

Table 5 Participants in the *Share Multiple* condition borrowed more features from their partners than other conditions

	Share Multiple	Share Best	Share One
Word phrases	15	9	6
Background color	3	0	2
Images	10	6	7
Layout	3	2	3
Surface patterns	1	1	1
Total	**32**	**18**	**19**

8   Group Rapport Rose for Partners Who Shared Multiple

A one-way ANOVA showed the group rapport differed significantly across conditions (F(2,83) = 4.147, p < 0.05). Tukey post-hoc comparisons of the three conditions indicate that group rapport increased in the *Share Multiple* condition ($\mu = 0.89$, SD = 3.06) compared to the others (p < 0.05). In absolute terms, rapport *only* increased in the Share Multiple condition (see Table 4).

9   *Share Multiple* Partners Took More Conversational Turns

An video analysis of speech duration during the group discussion showed that participants in the *Share Multiple* condition had significantly more frequent verbal exchanges (a higher number of speaker turns per minute of speaking time) than other conditions (F(2,39) = 3.506, p < 0.05). *Share Multiple* pairs averaged 12.1 (SD = 4.99) turns per minute, compared to 9.1 (SD = 2.62) and 8.6 (SD = 2.86) turns per minute, for *Share Best* and *Share One*, respectively. There were no significant between-condition differences for total number of speaker turns (F(2,39) = 0.695, p > 0.05), total speaking time (F(2,39) = 1.057, p > 0.05), or the ratio of high and low talkers (F(2,39) = 0.092, p > 0.05).

10   *Share Multiple* Pairs Borrowed More Features

In total, *Share Multiple* partners borrowed 32 features, *Share Best* 18, and *Share One* 19 (see Table 5). (The theoretical maximum for each condition is 140: 28 participants, 5 categories.) Participants in the *Share Multiple* condition borrowed significantly more features overall ($\chi^2 = 4.05$, p < 0.05).

Table 6 Pairs designs in the *Share Multiple* condition increased in similarity more than other conditions (std dev in parentheses)

	Before design critique	After design critique	Similarity shift
Share Multiple	2.59 (1.55)	3.50 (1.91)	0.91
Share Best	2.75 (1.71)	3.30 (1.97)	0.55
Share One	2.87 (1.81)	3.39 (1.85)	0.52

11 *Share Multiple* Pairs Reached a Better Consensus

Independent judges rated the partner ad similarity before and after the discussion. The similarity change provides a measure of shared perspective. Overall, final ads were more similar ($\mu = 3.40$, $SD = 1.91$) than initial ads ($\mu = 2.68$, $SD = 1.64$) ($t(3078) = 8.107$, $p < 0.05$). Tukey post-hoc comparisons of shifts by each pair show that similarity increased more for the *Share Multiple* condition (0.91) than the *Share Best* (0.55) or *Share One* conditions (0.52) ($p < 0.05$) (see Table 6).

12 Discussion

Sharing *multiple* designs led to several kinds of better outcomes. Simply *creating* multiple designs (without feedback) led to broader exploration, but not better results. The benefits were only realized if participants *shared* multiple designs. It's important to remember that participants worked on the same task for the same amount of time. The only variable was how many designs they created and shared. This section suggests reasons for why this simple act yielded differential outcomes, illustrating these with interview excerpts.

13 Hypotheses Revisited

13.1 Creating and Viewing Multiple Designs Leads to More Individual Exploration

The *Share Multiple* and *Share Best* participants explored significantly more broadly than the *Share One* participants. Creative work often benefits from broadly exploring possibilities before choosing a direction to refine (Buxton 2007). As this study and prior work found (Dow et al. 2010; Nielsen and Faber 1996), rapidly producing alternatives and getting feedback on them yields higher-quality, more-diverse results. As one *Share Multiple* participant described after seeing her partner's designs: "they were completely different from mine and I was like holy hell, that's pretty good. I didn't think about that." Another participant claimed that "getting a different perspective helped and also seeing different ideas—not flaws in mine, but different ideas in his that I'd like to borrow."

Experienced participants created less diverse designs than novices; their ads were also rated higher. Expert designers can rapidly construct entailments, mentally simulating design moves and their consequences (Schon 1995). Drawing on their many prior experiences, experts can often ignore or disregard the obviously bad options (Ericsson and Smith 1991). This foresight enables experts to strategically explore highly-promising subsets of the design space.

13.2 Sharing Multiple Designs Leads to More Productive Dialogue and Better Group Rapport

Across all conditions, there was a small but statistically significant decline in reported group rapport after the discussion (t(83) = 2.050, p < 0.05). While critique obviously provides a valuable channel for feedback and learning new information, negative critique can also degrade group relations and performance. Individuals' self efficacy may decrease in reaction to negative critique, which in turn lowers their performance (Dannels and Martin 2008). Furthermore, people receiving negative critique may resent the critique provider, poisoning team dynamics. In an understatement, one crest-fallen *Share Best* participant demurred, "she didn't make me feel like a total failure." Clearly the critique damaged her estimation of her abilities. Would this participant have felt better about herself and her team if she had shared multiple designs? In this study, group rapport actually increased when participants shared multiple designs. And prior work has shown that when teams generate lots of ideas, people feel more shared ownership and stronger team cohesiveness (Gerber 2010; Sutton and Hargadon 1996).

Pairs in the *Share Multiple* condition exchanged speaking turns significantly more often. As one *Share Multiple* participants said, "being able to see the other person's designs and actually bounce ideas back and forth... that helped clarify what was good design and what wasn't." Frequent exchanges helped participants discuss design tradeoffs and consider changes that address fundamental issues. As one participant said, "it got me thinking about who would click on an ad and why someone would click on an ad."

13.3 Sharing Multiple Designs Increases Conceptual Blending

Participants in the *Share Multiple* condition integrated more features and modified their designs to be more like their partner's. Participants often talked about the process of merging designs. One *Share Multiple* participant said, "we agreed we like elements of mine and I really like some elements of one of his and we just kind of did a mash-up and combined them." In contrast, a *Share Best* participant said, "we thought about some ideas, but we didn't really get to a consensus of what we were going to design." Likewise, a *Share One* participant said, "I didn't really get a lot of things to change on mine, so I just stuck with what I had." This notion of "sticking" with an idea did not surface in the *Share Multiple* condition.

Seeing multiple of a partner's designs provides more raw material for comparison. This was beneficial because comparison helps people understand underlying principles better than just one (Thompson et al. 2000). One participant in the *Share Multiple* condition said "(our) ads look different, but I feel like in general it's the same message that's getting across." Forming a stronger understanding of their partner's design rationale may be one reason *Share Multiple* participants reached more consensus and produced better results. Consensus is importantly different than groupthink, where a group blindly follows along one path without considering alternatives. In this study, convergence between Share Multiple pairs occurred *after* participants had explored many concepts.

14   Applying a *Share Multiple* Approach

Several interesting questions emerge about applying these results.

14.1   What Are This Study's Implications for Teaching and Practice?

Design organizations and educators can structure group work around creating and sharing alternative designs. For example, Stanford's introductory HCI course revised its curriculum to more strongly emphasize creating and comparing alternatives (http://cs147.stanford.edu). For designers and educators who already employ a "share multiple" approach, this result provides them empirical support.

14.2   (How) Can the Share Multiple Strategy Help with More Complex Design Work?

In this study, participants were able to create complete designs in a short amount of time. In many domains, sketches can be produced quickly, but creating *complete* designs is costly and time consuming. Does that mean for complex domains, a share multiple strategy can only be employed early in the design process? When creating multiple comprehensive designs is impractical, designers can still prototype and share alternatives to sub-problems. For example, in Web design it may be infeasible to produce three very different functional sites, but invaluable to create and test strategically selected elements. In fact, some of the world's best sites do so every day (Kohavi and Longbotham 2007).

15   Considerations for Using Ad Analytics in Experiments

This paper's advertising paradigm provides experimenters leverage when studying creative work. It offers strong quantitative benchmarks through its Web analytics, captures the views of many stakeholders, and provides measures of several different types of outcomes. While using advertising analytics is appealing for its ecological validity, this section shares three practical challenges.

15.1   Ensure Ads Are Shown Evenly

To maximize profit, many advertisers show ads differentially. If an ad performs well, it is shown more. If initial performance is poor, it is shown less (or not at all). Some platforms provide a setting to show ads more evenly; use this when available. Even with this setting, rotation may not be completely even; monitor this daily. It is important to run a pilot test with a few ads to determine effective keywords, budget (cost per click and daily maximums). Specifying geographical regions and time of day can also help generate a sufficient volume of impression and clicks.

15.2   Be Mindful of Ad Market Capacity

While it may appear that—given a sufficient budget—advertising platforms have unlimited capacity, this is not the case. Web adverting succeeds because it shows ads relevant to a user's current interests (http://en.wikipedia.org/wiki/Online_advertising). Showing irrelevant ads yields few clicks and little revenue for the host. On a given day, a finite number of people search for a particular topic like AIDS, design, or real-estate. That's the upper bound of relevant ads that can be served.

It's preferable to show all ads simultaneously to factor out differential effects of external variables, such as day of week, time of year, current events, etc. Furthermore, one needs a sufficient number of impressions and clicks to make meaningful statistical distinctions. For this paper's study, 80 ads pushed the limits of how many alternate ads can be simultaneously shown. Circulating more ads would have sliced the market of available impressions too thin.

15.3   Gather Multiple Outcome Measures

Which measure is "best"? In our study, online click performance did not correlate with overall rating. Some unattractive ads receive many clicks; the Web has a preponderance of such examples. Is an advertisement's success defined by its click-through performance or by expert ratings? Each tells a valuable story.

16   Conclusions and Future Work

This paper found that when people produce and share multiple alternatives with peers, they explore more diverse ideas, integrate more of their partner's features, engage in more productive design conversations, and ultimately, create higher-quality work. Many designers already practice this approach. These results suggest that more practitioners and teachers might beneficially adopt a "share multiple" strategy. More broadly, this work raises several important questions.

First, (how) do these results become general when applied to different types of groups? In this study, participants were independently recruited with no prior collaboration. In most professional work, collaboration is longitudinal, and power relationships and social dynamics are more complex. In this study, all group members performed the same role. Often for a variety of reasons, different team members perform different functions. Cross-functional teams can add value in both professional and learning contexts, such as jigsaw learning where different students are responsible for complementary parts of a topic (Aronson et al. 1978). What does sharing multiple designs mean for cross-functional teams and how does the outcome change depending on who does the creating and sharing?

Second, recent research on "the crowd within" suggests that at least some of the benefits of aggregating many people's perspectives can be accomplished by providing individuals with a structured approach to considering alternatives (Herzog and Hertwig 2009). This study witnessed several benefits of group discussion; can structured reflection help individuals benefit similarly?

Third, we hypothesize the share multiple condition benefited in two ways. *Creating* several alternatives spread participants' investment and discouraged fixation. *Seeing* others' designs gave participants a larger palette to work from. An important step for future work is to separate these two effects. One strategy would be to have designers supplement their own creations with previously created examples. An alternative would be for software to synthesize design alternatives (Hartmann et al. 2008; Lee et al. 2010).

Fourth, the benefits of rapidly creating and sharing multiple alternatives are myriad. How might software tools help designers explore more broadly? Initial results are promising (Hartmann et al. 2008; Lee et al. 2010); more exciting work remains.

Appendix A: Graphic Design Assessment

Instructions: For each of the statements below, indicate (True or False) whether or not the statement is a rule of graphic design	
1	Mix serif and sans serif fonts in order to give variety to the ad	F
2	To help balance the ad, leave slightly more space at the top relative to the bottom of the ad	
3	Create a visual separation between the text and the background	T
4	Angle the text in order to contrast different parts of the ad	F
5	Keep all elements in the ad aligned to one side	F
6	Create multiple visual focal points in order to attract attention to the ad as a whole	F
7	Use borders or white around text and images to help frame the content	T
8	You may use repetition to create a consistent and balanced look	T
9	You may break alignment to draw the viewer's attention to important elements in the ad	T
10	Draw the viewer's attention to important elements by contrasting scale	T

Appendix B: Advertising Design Brief

Assignment

You have been hired to design a graphic advertisement for FACEAIDS.org. You will learn to use a new graphic design tool, design provisional ads, and create a final ad to be posted through the Google ad network.

Goals

Keep in mind the following goals as you create your ads:

(a) Increase traffic to the *FaceAIDS* website: http://faceaids.org/
(b) Reach out to the target audience: students interested in improving global healthcare equality and making a difference in the AIDS epidemic in Africa.
(c) Impress the clients from *FaceAIDS*, who will rate your ads. The client wants an ad that fits their overall aesthetic and theme (see below).
(d) Create ads with effective graphic design. Ad professionals will rate your ads.

What is FaceAIDS?

FaceAIDS is a nonprofit organization dedicated to mobilizing and inspiring students to fight AIDS in Africa. *FaceAIDS* aims to build a broad-based movement of students seeking to increase global health equality. The organization raises awareness and funds, with the goal of increasing global health equality starting with the AIDS epidemic in Africa.

Theme and Aesthetic for the FaceAids Ad

FaceAIDS would like an advertisement that embodies the theme and general aesthetic of the organization. In particular, they are looking to encourage high school and college students interested in getting involved in service or social justice work to start *FaceAIDS* chapters on their campuses, as a leadership development opportunity and a way to join a vibrant, impactful community of like-minded, driven peers. In general they are looking for an ad that is tasteful, creative, professional, visually appealing, and conveys a clear message about the organization.

Rules/Requirements

– You may download and use graphics & images as you see fit.
– You may not use another company's logo, copyrighted images, profanity, obscenity or nudity. Unacceptable ads will be rejected by the research team.

Do not include the magazine's URL on the ad. Clicking the ad will direct the user to the site.

References

Arkes HR, Blumer C (1985) The psychology of sunk cost. Organ Behav Hum Decis Process 35 (1):124–140

Aron A, Aron EN, Smollan D (1992) Inclusion of other in the self scale and the structure of interpersonal closeness. J Pers Soc Psychol 63(4):596–612

Aronson E, Bridgeman D, Geffner R (1978) Interdependent interactions and prosocial behavior. J Res Dev Educ 12(1):16–27

Ball LJ, Ormerod TC (1995) Structured and opportunistic processing in design: a critical discussion. Int J Hum-Comput Stud 43(1):131–151

Bao P, Gerber E, Gergle D, Hoffman D (2010) Momentum: getting and staying on topic during a brainstorm. In: Proceedings of conference on human factors in computing systems, ACM, New York, pp 1233–1236

Ben-David I, Graham JR, Harvey CR (2007) Managerial overconfidence and corporate policies. National Bureau of Economic Research working paper series no. 13711

Brandt J, Dontcheva M, Weskamp M, Klemmer SR (2010) Example-centric programming: integrating web search into the development environment. In: Proceedings of conference on human factors in computing systems, ACM, New York, pp 513–522

Brereton M, Cannon M, Mabogunje A, Leifer L, Brereton M, Cannon M, Mabogunje A, Leifer L (1996) Collaboration in design teams: how social interaction shapes the product. In: Analyzing design activity. Wiley, Chichester

Buxton B (2007) Sketching user experiences: getting the design right and the right design. Morgan Kaufmann, Amsterdam

Cross N (2004) Expertise in design: an overview. Des Stud 25(5):427–441

Curhan JR, Elfenbein HA, Xu H (2006) What do people value when they negotiate? Mapping the domain of subjective value in negotiation. J Pers Soc Psychol 91(3):493–512

Dannels DP, Martin KN (2008) Critiquing critiques: a genre analysis of feedback across novice to expert design studios. J Bus Tech Commun 22(2):135–159

Davidoff S, Lee MK, Dey AK, Zimmerman J (2007) Rapidly exploring application design through speed dating. In: Proceedings of conference on ubiquitous computing, Innsbruck

Diehl M, Stroebe W (1987) Productivity loss in brainstorming groups: toward the solution of a riddle. J Pers Soc Psychol 53(3):497–509

Dow SP, Heddleston K, Klemmer SR (2009) The efficacy of prototyping under time constraints. In: Proceedings of ACM conference on creativity and cognition, ACM, New York, pp 165–174

Dow S, Glassco A, Kass J, Schwarz M, Schwartz DL, Klemmer SR (2010) Parallel prototyping leads to better design results, more divergence, and increased self-efficacy. Trans Comput-Hum Int, Article 18, 17(4):24

Dweck C (2007) Mindset: the new psychology of success. Ballantine Books, New York

Ericsson KA, Smith J (1991) Toward a general theory of expertise: prospects and limits. Cambridge University Press, Cambridge

Fauconnier G, Turner M (2003) The way we think: conceptual blending and the mind's hidden complexities. Basic Books, New York

Felps W, Mitchell T, Byington E (2006) How, when, and why bad apples spoil the barrel: negative group members and dysfunctional groups. Res Organ Behav 27:175–222

Finke RA, Ward TB, Smith SM (1996) Creative cognition: theory, research, and applications. The MIT Press, Cambridge, MA

Gaver WW, Beaver J, Benford S (2003) Ambiguity as a resource for design. In: Proceedings of the SIGCHI conference on human factors in computing systems, ACM, New York, pp 233–240

Gerber E (2010) Prototyping practice in context: the psychological experience in a high tech firm. J Des Stud

Hampton JA (1987) Inheritance of attributes in natural concept conjunctions. Mem Cognit 15(1):55–71

Hartmann B, Yu L, Allison A, Yang Y, Klemmer SR (2008) Design as exploration: creating interface alternatives through parallel authoring and runtime tuning. In: Proceedings of the conference on user interface software and technology, ACM, New York, pp 91–100

Herzog SM, Hertwig R (2009) The wisdom of many in one mind. Psychol Sci 20(2):231–237

Hyland F, Hyland K (2001) Sugaring the pill: praise and criticism in written feedback. J Second Lang Writ 10(3):185–212

Iyengar SS, Lepper MR (2000) When choice is demotivating: can one desire too much of a good thing? J Pers Soc Psychol 79(6):995–1006

Janis IL (1982) Groupthink: psychological studies of policy decisions and fiascoes. Wadsworth, New York

Jansson D, Smith S (1991) Design fixation. Des Stud 12(1):3–11

Kershaw TC, Ohlsson S (2004) Multiple causes of difficulty in insight: the case of the nine-dot problem. J Exp Psychol Learn Mem Cogn 30(1):3–13

Kohavi R, Longbotham R (2007) Online experiments: lessons learned. Computer 40:103–105

Kosara R (2007) Visualization criticism – the missing link between information visualization and art. In: Proceedings of the conference on information visualization. IEEE Computer Society, Washington, DC, pp 631–636

Larrick RP (2009) Broaden the decision frame to make effective decisions. In: Locke E (ed) Handbook of principles of organizational behavior. Wiley, Chichester, UK

Lee B, Srivastava S, Kumar R, Brafman R, Klemmer SR (2010) Designing with interactive example galleries. In: Proceedings of the conference on human factors in computing systems, ACM, New York, pp 2257–2266

Leifer L (2010) Dancing with ambiguity: design thinking in theory and practice. http://hci.stanford.edu/courses/cs547/speaker.php?date=2010-04-09

Mark G, Gonzalez VM, Harris J (2005) No task left behind?: examining the nature of fragmented work. In: Proceedings of the conference on Human factors in computing systems, Portland, pp 321–330

Marsh RL, Landau JD, Hicks JL (1996) How examples may (and may not) constrain creativity. Mem Cognit 24(5):669–680

Moran TP, Carroll JM (1996) Design rationale: concepts, techniques, and use. CRC Press, Mahwah, NJ

Nickerson RS (1998) Confirmation bias: a ubiquitous phenomenon in many guises. Rev Gen Psychol 2:175–220

Nielsen J, Faber JM (1996) Improving system usability through parallel design. Computer 29(2):29–35

Ranganath R, Jurafsky D, McFarland D (2009) It's not you, it's me: detecting flirting and its misperception in speed-dates. In: Proceedings of conference on empirical methods in natural language processing, Association for Computational Linguistics, pp 334–342

Schon DA (1995) The reflective practitioner: how professionals think in action. Ashgate, Aldershot

Schrage M (1999) Serious play: how the world's best companies simulate to innovate. Harvard Business School Press, Boston

Schwartz DL (1995) The emergence of abstract representations in Dyad problem solving. J Learn Sci 4(3):321

Schwartz B (2004) The paradox of choice: why more is less. Ecco, New York

Smith S (1993) Constraining effects of examples in a creative generation task. Mem Cognit 21:837–845

Stroebe W, Diehl M (1994) Why groups are less effective than their members: on productivity losses in idea-generating groups. Eur Rev Soc Psychol 5:271

Sutton R, Hargadon A (1996) Brainstorming groups in context: effectiveness in a product design firm. Adm Sci Q 41:685

Taylor D, Berry P, Block C (1958) Does group participation when using brainstorming facilitate or inhibit creative thinking? Adm Sci Q 3(1):23–47

Thomke S, Nimgade A (2000) IDEO product development. Harvard Business School Case, Boston

Thompson L, Gentner D, Loewenstein J (2000) Avoiding missed opportunities in managerial life: analogical training more powerful than individual case training. Organ Behav Hum Decis Process 82(1):60–75

Tohidi M, Buxton W, Baecker R, Sellen A (2006) Getting the right design and the design right. In: Proceedings of the SIGCHI conference on human factors in computing systems, ACM, New York, pp 1243–1252

Warr A, O'Neill E (2005) Understanding design as a social creative process. In: Proceedings of the conference on creativity & cognition, ACM, New York, pp 118–127

Wisniewski E, Gentner D (1991) On the combinatorial semantics of noun pairs: {minor} and major adjustments to meaning. In: Understanding word and sentence. North Holland, Amsterdam, pp 241–284

Zwicky F (1969) Discovery, invention, research through the morphological approach. MacMillan, New York

Towards a Paradigm Shift in Education Practice: Developing Twenty-First Century Skills with Design Thinking

Christine Noweski, Andrea Scheer, Nadja Büttner, Julia von Thienen, Johannes Erdmann, and Christoph Meinel

1 What Is the Intention of This Article?

Science, business and social organizations alike describe a strong need for a set of skills and competencies, often referred to as twenty-first century skills and competencies (e.g. Pink, Wagner, Gardner). For many young people, schools are the only place where such competencies and skills can be learned. Therefore, educational systems are coming more and more under pressure to provide students with the social values and attitudes as well as with the constructive experiences they need, to benefit from the opportunities and contribute actively to the new spaces of social life and work. Contrary to this demand, the American as well as the German school system has a strong focus on cognitive skills, acknowledging the new need, but not supporting it in practice. Why is this so? True, we are talking about a complex challenge, but when one makes the effort to take a closer look, it quickly becomes apparent that most states have not even bothered to properly identify and conceptualize the set of skills and competencies they require. Neither have they incorporated them into their educational standards.

No wonder, teachers stand helpless in the face of new challenges and have – more or less – only their personal experience and good will to fall back on. An approach which is naturally not successful on a broad scale.

Developments in society and economy require that educational systems equip young people with new skills and competencies, which allow them to benefit from the emerging new forms of socialization and to contribute actively to economic development under a system where the main asset is knowledge (Ananiadou and Claro 2009, p. 5).

C. Noweski • A. Scheer • N. Büttner • J. von Thienen • J. Erdmann
Hasso-Plattner-Institute, Potsdam, Germany

C. Meinel (✉)
Hasso-Plattner-Institut (HPI), für Softwaresystemtechnik GmbH, Prof.-Dr.-Helmert-Str. 2-3, Potsdam 14482, Germany
e-mail: meinel@hpi.uni-potsdam.de

H. Plattner et al. (eds.), *Design Thinking Research*, Understanding Innovation, DOI 10.1007/978-3-642-31991-4_5, © Springer-Verlag Berlin Heidelberg 2012

The research team e.valuate has worked on this challenge for 1 year now and wants to share some of its findings.

We will start by introducing skills and competencies behind the term twenty-first century skills, as well as the concept of constructivist teaching and learning – a methodology most promising to cope with the new demands. We will then explain, why Design Thinking, understood as constructivist methodology, is especially appropriate to enable teachers to prepare our students to cope with the challenges of the twenty-first century. In the fourth part, we will introduce an empirical study undertaken to prove the hypotheses derived from part three.

2 What Are Twenty-First Century Skills and Why Is Everybody Talking About Them?

Initiatives on the teaching and assessment of twenty-first century skills originate in the widely-held belief shared by several interest groups – teachers, educationalists, policy makers, politicians and employers – that the current century will demand a very different set of skills and competencies from people in order for them to cope with the challenges of life as citizens, at work and in their leisure time (e.g. Pink 2006; Wagner 2010; Gardner 2007). Initiatives such as the Partnership for twenty-first century skills and the Cisco/Intel/Microsoft assessment and teaching of twenty-first century skills project also point to the importance currently attached to this area not only by researchers, practitioners and policy makers but also the private sector. Supporters and advocates of the twenty-first century skills movement argue for the need for reforms in schools and education to respond to the social and economic needs of students and societies in the twenty-first century. Most of them are related to knowledge management, which includes processes related to information selection, acquisition, integration, analysis and sharing in socially networked environments.

Before presenting which skills and competencies are broadly understood in this context, we would like to define the terms "skills" and "competence" and make clear, how they relate to each other.[1]

---

[1] One useful distinction between the two is provided by the OECD's DeSeCo project: A competence is more than just knowledge or skills. It involves the ability to meet complex demands, by drawing on and mobilizing psychosocial resources (including skills and attitudes) in a particular context. For example, the ability to communicate effectively is a competence that may draw on an individual's knowledge of language, practical IT skills and attitudes towards those with whom he or she is communicating (Rychen and Salganik 2003). The European Commission's Cedefop glossary defines the two terms as follows: A skill is the ability to perform tasks and solve problems, while a competence is the ability to apply learning outcomes adequately in a defined context (education, work, personal or professional development). A competence is not limited to cognitive elements (involving the use of theory, concepts or tacit knowledge); it also encompasses functional aspects (involving technical skills) as well as interpersonal attributes (e.g. social or organizational skills) and ethical values (Cedefop 2008).

A competence is thereby a broader concept that may actually comprise skills (as well as attitudes, knowledge, etc.). However, the terms are sometimes used interchangeably or with slightly different definitions in different countries and languages. This should always be kept in mind.

Based on the above, we will stick to the OECD working definition of twenty-first century skills and competencies: *Those skills and competencies young people will be required to have in order to be effective workers and citizens in the knowledge society of the twenty-first century.*[2]

A multitude of authors have laid down their concepts of twenty-first century skills. Giving a broad view of society, we want to present three. Researcher, author and internationally acclaimed speaker Tony Wagner (former teacher and principal) calls twenty-first century skills the *seven survival skills for careers, college and citizenship* (Wagner 2011) and distinguishes in his book *The global achievement gap*:

- Critical thinking and problem solving
- Collaboration across networks and leading by influence
- Agility and adaptability
- Initiative and entrepreneurialism
- Effective oral and written communication
- Accessing and analyzing information
- Curiosity and imagination.

Successful author and connoisseur of American politics Daniel Pink describes in his book *A Whole New Mind* six essential aptitudes: *on which professional success and personal fulfillment nowadays depend.* He distinguishes:

- Design: to detect patterns and opportunities
- Story: to create artistic and emotional beauty and to craft a satisfying narrative
- Synthesis: to combine seemingly unrelated ideas into something new
- Empathy: ability to empathize with others and to understand the subtleties of human interaction
- Meaning: to find joy in one's self and to elicit it in others and to stretch beyond the quotidian in the pursuit of purpose and meaning.

Harvard professor Howard Gardner builds on decades of cognitive research and rich examples from history, politics, business, science, and the arts when he describes: *the specific cognitive abilities that will be sought and cultivated by leaders in the years ahead* in his book *Five Minds for the Future*. The five Minds are:

- The Disciplinary Mind: the mastery of major schools of thought, including science, mathematics, and history, and of at least one professional craft.
- The Synthesizing Mind: the ability to integrate ideas from different disciplines or spheres into a coherent whole and to communicate that integration to others.

---

[2] Ananiadou and Claro (2009).

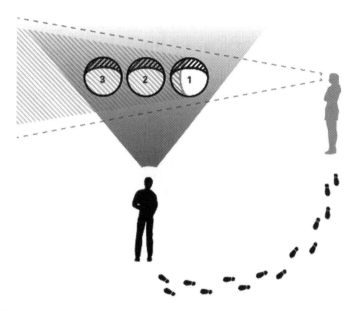

Fig. 1 Changing perspective (By Christine Noweski and Elias Barrasch 2011)

- The Creating Mind: the capacity to uncover and clarify new problems, questions and phenomena.
- The Respectful Mind: awareness of and appreciation for differences among human beings and human groups.
- The Ethical Mind: fulfillment of one's responsibilities as a worker and as a citizen.

We decided to offer these lists here to give you, our dear reader, a look at the broadness of the discussion. According to context, audience and goal, the descriptions vary a lot, but center around the same basic concepts.

After analyzing and comparing many more approaches, we decided to work with the rather abstract psychological three-tier categorization of competences offered by Himmelmann (2005). He classifies key competences into:

- Cognitive abilities (Fig. 1:1)
- Affective, moral attitudes (Fig. 1:2)
- Practical, instrumental skills (Fig. 1:3).

Figure 1 shows how dangerous it becomes when one is focusing too much on only one, as is today the case with cognitive abilities. It may overshadow the other competencies completely. What is happening in schools right now (and this includes both the very different American and German school systems) is a strong emphasis on measuring and comparing cognitive abilities. This is supported by multiple guidelines for teachers, as well as students, on how to find one's way through this system.

The goal of the years to come has to be to find a way back to a perspective, where teachers envision all three categories to lay down the base for twenty-first century skilled students. It's a challenge to confront the personal desire to do things the way one has always done: it feels so safe and good, why should one put this at risk? To reach out into the unknown is uncomfortable for most people, so why should teachers feel any differently about this? The few who still seek to try the shift, are often hindered by bureaucratic structures and hierarchies that are built on old principles. More on this in chapter "The Faith-Factor in Design Thinking: Creative Confidence Through d.school Education?", when we will describe the ideal role of a teacher in constructivism and how it differs from reality.

3 Opportunity Constructivism: What? Why? and How?

In this chapter, we will introduce the theory of constructivism and its implications on learning and teaching, in order to gain an overall understanding before describing in more detail the problem solving approach of Dewey and the Project Method Kilpatrick that can be seen as a still used predecessor of Design Thinking. We will then describe Design Thinking as a teaching and learning methodology while focusing on its potential to mediate twenty-first century skills.

There are three main philosophical frameworks under which learning theories fall (see Fig. 2): behaviorism, cognitivism, and constructivism. Behaviorism focuses on objectively observable aspects of learning. Cognitive theories look beyond behavior to explain brain-based learning. Both can be considered as approaches of realism (more on realism see e.g. Miller 2010 and Zalta 2010).

Learning can also be understood from a constructivist perspective, in which *learning is a process of understanding, which leads to modifications in the behavior of the learner due to experiences,*[3] a process of individually self-organizing knowledge. Learning theories from Jean Piaget, Jerome Bruner, Lev Vygotsky and John Dewey serve as a basis for constructivist learning theory. Several authors need to be mentioned because constructivist theory is a broad approach towards learning. Shared convictions are that the process of learning is unpredictable and knowledge constantly altered through new insights, which are gained through individual experiences (Reich 2008; Kolb 1984). In realism the learner is regarded as an independent observer of objects. In contrast, constructivism integrates the learner within his own observations in a cycle of creation and observation. An interactive relation between the observer and the observed arises (for an easier understanding see Fig. 3).

The educationalist and philosopher John Dewey regarded the interaction between the subject and the world as essential for gaining knowledge. Dewey's understanding identified learning as a direct process of the structured interaction of

---

[3] Hasselhorn and Gold (2006, p. 35).

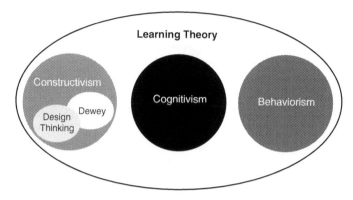

Fig. 2 Philosophical frameworks of learning theory (By Christine Noweski 2011)

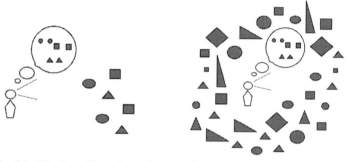

Empirical Realism: Observing a given reality independent from oneself.

Constructionism: The observer as part of his environment and the reality he chooses.

Fig. 3 The learner and his environment (By Andrea Scheer 2011)

humans and their natural and social environment. These interactions produce experiences which modify further interaction – then, learning takes place (see Hasselhorn and Gold in the beginning of this chapter).

There is no me without us.[4] Perception and knowledge is only developed in relation to and through interaction with the object and its context. Therefore, learning in the constructivist perspective is a process of constantly adapting to situations, which consist of ever-changing relations between subject, object and context. Navigating through this process and identifying relations creates knowledge. However, constructivism is neither a method nor a universal model, but it defines the perspectives on learning and knowledge.

---

[4] Dewey (1931, p. 91).

Education today is focused on breaking down complex phenomena into abstract parts (e.g. subjects, different topics within subjects). Aspects of knowledge are considered in their singularity, and distributed inductively[5] to the student. It is easier to only look at the parts and pieces of a clock than figuring out its complex correlations. Still, the clock only makes sense as whole, and the pieces need to be properly reassembled into the complexity of relations between its components. The process of reassembling pieces of knowledge into the complex phenomena is seldom realized in schools today. This makes it hard for student to see links between the subjects and topics to be learned in school and the real-life context. It is hard for the teacher to realize complex deductive[6] learning, as learning methods and theories are still very abstract. But, how do we make complex phenomena understandable without breaking them down into too many abstract parts?

3.1 What a Constructivist Learning Design and Teaching Should Look Like

Pedagogical science states that the competences claimed in chapter "Design Thinking Research" can be taught especially well through a deductive method from the perspective of constructivist learning (Weinert 2003; Knoll 1991; Reich 2008). Constructivism as described above looks at complex phenomena as a whole within its context and from the perspective of the observer.

Dewey stated the following three aspects as essential for a convenient learning design:

- Involvement of students
- Available space for experiencing
- Deductive instruction and
- Possibilities for construction.

Here, the teacher acts as a mediator between the different entities, and defines how the students go through their individual process of understanding. The teacher has a manipulative function laying down the framework for subject, object and context. The teacher as a facilitator of learning should consequently be able to design learning experiences. As participation and engagement of the students are crucial characteristics of constructivist learning (Reich 2008), the teacher should involve students in the learning design, e.g. by looking at students' interests when developing a problem statement or project challenge. Furthermore, teachers need to give space to the students to try out different mental models and methods. The

---

[5] Inductive as defined by the Oxford Dictionary: "inference of general laws from particular instances".

[6] Deductive as defined by the Oxford Dictionary: "inference of particular instances from a general law."

students would then have the opportunity to connect abstract knowledge with concrete applications and thereby be able to convert and apply abstract and general principles (instructions) in meaningful and responsible actions in life (construction). In a nutshell, a good lesson design needs to be a balanced composition of instruction and construction, or as Dewey would say construction through instruction (Dewey 1931; Knoll 1991). It should consist of a plan of how students can experience certain situations and how teachers can enable and support this experience. A good learning design is what schools have usually failed to provide up until today. The HOW, e.g. the instruction to execute constructivist learning, is either missing (free construction only) or too inductive (instructed construction only). It is an art to find the right balance between giving a frame through instruction and offering freedom for construction through paths within this frame – it is the art of teaching.

Teacher education should meet these implications by preparing the teacher not only in subject content, but also in meta-competencies like facilitation and design of learning experiences.

3.2 Abstract Concept: Project-Method Based on John Dewey

Dewey addressed the question of teaching complex phenomena as a whole by proposing recommendations for constructivist problem-solving, which was later transformed into the Project-Method by his student William Heard Kilpatrick in 1918. Dewey's approach was related to the natural sciences in that it started with an inquiry unfolding a problem or difficulty, which was then the motivation for further analyses and exploration. New insights are the basis for an explanation of that inquiry, and followed by a plan of action to solve the problem. Dewey recommended considering the following aspects:

- Problems situated in a real-life context
- Interaction of thinking and action
- Interaction and sharing of knowledge between learner and teacher
- Problem-solving and interpretation of insights
- Reflecting and understanding through application of ideas.

In conclusion, Dewey's perspective on learning and education is centered around a real-life inquiry, which has to be analyzed as a complex whole (deductive). The inquiry acts like *a magnet for further analysis of content and input of several disciplines in order to explain and solve that complex inquiry as a whole.*[7]

Dewey's recommendations have been around for more than a century, and although there is a common wish for their implementation, they are seldom practiced in schools. We believe this is because his theory is too abstract, and therefore hard for teachers to practically implement in the classroom. That is why

[7] Dewey (1931, p. 87).

we compared the realization of Deweys recommendations and its adaption in the Project-Method by Kilpatrick with the Design Thinking method.

We believe that Design Thinking builds on Dewey's argument of complex inquiry-based learning, and that it gives concrete recommendations for distributing a complex phenomenon without breaking it down and diluting the relations between subject, object and context, at the same time being digestible for the student and implementable for the teacher.

3.3 Concrete Framework: Design Thinking

In this paragraph, we will describe our understanding of the concept, and the methods employed in Design Thinking. As there is nothing such as an agreed theory in this field, we stick to our experience, observations and insights from expert interviews.

Design Thinking conveys a thinking and working style of its own uniqueness, while employing existing methods and theoretical concepts. The concept offers a frame to work on solving complex challenges, which Rittel (1973) described as *wicked problems* (1973). It also provides a pathway for innovations by creating and iterating inventions. Due to its innovation stimulating character, it has gained increasing attention and relevance over the last decades, especially in recent business practice (Amabile 2008; Runco 2004).

Building on the theoretical concepts of Dewey, Peirce and others, Design Thinking reproduces knowledge through action with the goal of changing existing situations into preferred ones. These challenges are tackled in interdisciplinary teams with a clear focus. The teams should ideally work together in a flexible working environment and in creative freedom, while at the same time being guided systematically through an iterative process. A coach mentors the team with methodological experience. There should be an emotional distance between the team member and coach, while at the same time sufficient closeness to always know when intervention is needed.

Throughout this process, all actions are aligned to a certain target, mostly the user for whom the project is designed. All of this together distinguishes the design thinking approach from usual business or technology driven concepts (see Fig. 4). Nonetheless, design thinking methodology acknowledges both of these concepts and tries to integrate these perspectives. It does this by transferring trends from science and practice, not forgetting that a holistic and fruitful innovation catalyzes human needs, technological feasibility and economic viability (Brown 2008).

Fig. 4 Design Thinking
approach (Based on Tim
Brown 2006)

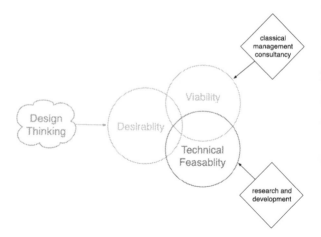

3.4   Excursus: Why Twenty-First Century Societies Need Innovation and Future Innovators Need Twenty-First Century Skills

With everyday complexity increasing, political concerns more intertwining, technologies changing faster, product cycles getting shorter and economic competition tightening, innovative capacities have become crucial to survive in a changing society and work life as a state, a company, and an individual (for further reading, see Freeman and Soete 1997). Without innovation there is no progress and without creative, skilled people who can meet these future demands there is no innovation. That's why future innovators, as social as well as a professional people, need to be equipped with twenty-first century skills (Carroll et al. 2010).

An innovation, in contrast to an invention, is not merely the addition of something new or the creation of an idea but a newness that provokes and instigates a economic, social and technical change through its realization and application. This is exemplified in the transformation from sketch into implementation (Fagerberg 2003; Schumpeter 1961). Though Design Thinking not (yet) solely regards the implementation part itself, it contributes to the innovation progress through its conceptual setting and by employing people with an innovative thinking and acting style. On the one hand, inventions are created by deploying an elaborate process with a user-centered approach and by merging people and knowledge from different expertise fields and disciplinary perspectives, knowing that most surprising innovations are often combinations and transformations from other already existing areas. On the other hand, Design Thinking encourages and develops a certain mindset in which we believe can accomplish the demand of the twenty-first century. A design thinker, for example, goes out into the field, holds dialogues with different stakeholders, observes (perhaps using cases and needs that are expressed indirectly) and immerses him or herself into another person's world. In this way, design thinkers also use all their analytical as well as their creative senses and abilities.

In the 1980s Drucker (1985, p. 72) described successful innovators as being *conceptual and perceptual* and using *both the right and left sides of their brains.* This is an individual who has expert knowledge in a special field and an inventive talent, a person who is conscious and assiduous, devoted and engaged, untiring and driven by learning from failures. Interestingly, these innovative qualities perfectly capture the personality and mindset of a design thinker. They are applied throughout Design Thinking, as we were able to prove in the empirical survey described in the next chapter.

3.5 How Does Design Thinking Work?

In this passage, we will briefly describe the above-mentioned systematic. The method of Design Thinking merges successful models from psychology, economics and pedagogy. Designers have intuitively applied them over a long period of time and, since the 1960s, reflectively and systematically put them together into an educational concept that also allows novices to work with a process that provides them with orientation and stability. Every step in the process thereby mirrors a particular attitude of the designerly way of thinking. Moreover, design thinkers are provided with experience about difficulties and obstacles of team dynamics in the corresponding phase.

3.5.1 Board the Journey

In literature and practice, various process models exist, with process phases differing and their naming varying. Leaning on Erdmann's circular model (see Fig. 5), we comprehend the design thinking process featuring the phases Understand, Observe, Synthesis, Ideate, Prototype and TEST.

Basically, the process follows these six steps that build on each other while preserving a cyclical and iterative nature. The star's outer lines and imagined arrows illustrate that it is possible and desired to move from one phase to any other at any point of time, as well as to repeat the whole process or just certain stages. In conclusion, there are multiple itemizations of each phase that derive from free iterations of itself.

In each phase, the most important results are the insights about users or ideas deriving from these that have the goal of solving inconveniences. These are then cumulated and documented in the star's inner centre. Thanks to the iterative approach, they can be looked up and modified again at any time in the ongoing process. This is very useful to integrate crucial insights into earlier findings and to generate new insights out of earlier ones. Each step in the process is limited in time and interim presentations of the status quo are utilized as demanding landmarks along the process.

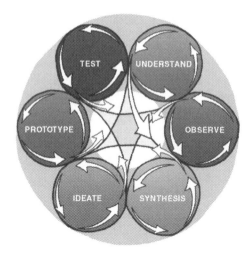

Fig. 5 Design Thinking process (After Johannes Erdmann 2010)

3.5.2 Where to Start

In the first step of the process "Understand" the initial task means to discuss in a team and work on a shared understanding of the challenge regarding its context, and dependencies. Successful design thinking teams often spend most of their project time exploring and understanding. First come the challenge and the user, later the possible solution spaces. Only by spending lots of time in this early phase, can a user-focused solution be ensured later on.

Furthermore, an agreed-on challenge helps the individuals to grow together as a team and make sure, everybody's knowledge, perspectives and skills can be utilized in the process. For more information, see the excellent article by Paulus (2000).

3.5.3 Be an Explorer

The aim of the next phase "Observe" is to get a 360°-overview about possible solution spaces. Besides interviews and observations it is often helpful for one to conduct the activities of the user him or herself, meaning to step into the role of the user and thereby to build up a special sense of empathy. For more information on methods in particular phases, we recommend checking out IDEO (2011).

In this phase, the team should take the time to look at as many different contexts as possible, because it often shows that interesting solutions in one particular challenge already exist in other contexts and can be successfully transferred.

3.5.4 Enter the Molten Bath

Experiences from the observe phase are exchanged in the "Synthesis" stage, where the most fruitful insights are compiled and distilled, eventually reframing the initial questions according to the findings. This is the first *moment of truth* where the team that enjoyed diverging over a broad mass of information, exploration and solution spaces has to now converge to a point of view that has the power to give them the necessary drive for the next diverging session (necessary energy for the next loop in Fig. 4). This is often the hardest milestone in the process and proceeds with a lot of discussion and an abrupt loss of motivation. Teams that manage to get out of this abstract bottleneck still united as a team, with a shared and clear understanding of the challenge to work on, are usually the ones that will succeed.

3.5.5 Embark on the Idea RoundAbout

In the IDEATE phase, solutions are generated individually and in the team by applying multiple forms of bodystorming,[8] including brainstorming, sketching, acting out use cases and rough prototypes. A set of rules helps to preserve a positive team dynamic and encourages building on the ideas of others as well as to encourage uncommon ideas. There are different definitions but you may want to check IDEO's collection at Open IDEO, which has a nice description of each single one. Thereafter, the most suspicious, promising ideas are chosen in the team (another point of converging).

3.5.6 Become a Master-Builder and Actor

In the next step of "Prototyping," selected ideas are made tangible. This can mean to build a model or to prepare a role play that lets an audience experience what the situation the team is working on feels like. There are two categories of prototypes: lookalikes and feelalikes. Prototypes don't have to be detailed nor perfect but should primarily deliver the main concept of the idea to outside people and answers to predefined questions to the team in order to prove and improve ideas and concepts. It is proven that the more crude a prototype is, the easier it is to gain conceptual feedback. Vice versa, the more refined the prototype is, the more detailed and focused on the appearance the feedback will be. For more information on this, see the excellent dissertation by Edelman (2011).

3.5.7 Proof of Concept

The team then presents the developed prototypes to designated users to let them try out and play with the idea. This TESTING aims to let crucial advantages and disadvantages become apparent through user feedback. In accordance to the

---

[8] Concept building on the common term brainstorming.

iterative principle, the team now is encouraged to go back to a previous phase and enhance or modify the idea or to start again from scratch.

3.6 How Does Design Thinking Contribute in Developing Twenty-First Century Skills?

In this passage, we outline which learnings and personality traits are fostered by Design Thinking and to what extent they contribute to Himmelmann's (2005) three-tier categorization of twenty-first century competences. Please be aware that these categories are overlapping and are categorized sequentially only for ease of understanding.

To operationalize Himmelmanns's abstract categories in our experiment, we used the ISK (ISK 2009). The ISK is a questionnaire that measures social competencies, subsumed under the four categories as shown in Fig. 6.

We will point out these competencies in design thinking process phases, where they are especially fostered, but as mentioned above, please be aware, that things go hand in hand at this level.

3.6.1 Cognitive Abilities

Learnings in this category comprise abilities regarding knowledge, cognizance, and comprehension.

In the OBSERVING phase design thinkers neutrally monitor people's actions in regard to what they say, how they act and what they actually mean. Information is generated and evaluated and divergent thinking is trained. While SYNTHESIZING, actions are mostly dedicated to cognitive skills: Information is selected and synthesized according to its relevance and degree of surprise. Convergent and abductive[9] procedures are also utilized.

Finding brainstorming questions requires different perspectives and phrasings. While IDEATING, divergent thinking and associative creativity come into play. Clustering ideas activates learning how to detect patterns and coherence by convergent thinking.

PROTOTYPING causes one to think about the details of the idea. Whereby, TESTING supports the ability to reflect upon one's own ideas, to cope with critical feedback and the comparison of expected and de facto performance. Convergent thinking is enhanced overall. Presenting findings at different milestones in front of a plenum, within strict time limitations, enhances the ability to put content in a nutshell while likewise conveying the message precisely to an audience.

---

[9] Described by Peirce as "guessing". The term refers to the process of arriving at an explanatory hypothesis (Peirce 1901, paragraph 219).

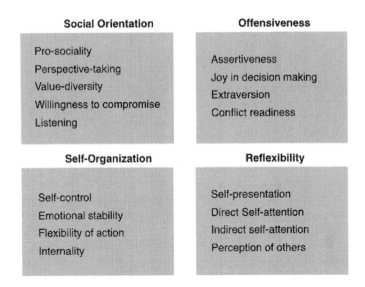

Fig. 6 Scales of the social competencies inventory (Based on Kanning 2009)

Communication throughout the process and set time frames serve to reflect in teams about content and non-content. Additionally, this helps to train the perception concerning oneself and others (direct self mindfulness, Indirect self mindfulness, perception of people).

3.6.2 Affective and Moral Attitudes

This rubric gathers all learnings concerning motivation, commitment, willing, attitudes, and habits.

In the first two phases UNDERSTAND and OBSERVE, prejudices and clichés are consciously avoided and dismantled by gaining a deep and broad understanding of the topic, people and context and also by exchanging different viewpoints within the team and with the outside, thereby learning to accept mindsets different than one's own (value pluralism and good listening). Further, the approach of getting involved in another person's thoughts and actions contributes well to the ability of one to empathize with others, and also the ability and willingness to socialize with and present oneself to unknown people (adoption of perspectives, extraversion, self-presentation).

In general, team communication and social skills regarding misunderstandings, opponent opinions, inner emotionality and rivalry between one's own and other preferences as well as actively finding a solution are challenged throughout all phases as interaction is demanded all the time in all directions. This category is operationalized by the ISK items: willingness to comprise, pro-sociality, willingness to deal with conflicts as well as emotional stability.

3.6.3 Practical and Instrumental Skills

In this category, learnings enfold abilities, proficiencies, and strategies.

Besides silent studying of situations, OBSERVING naturally requires talking to different stakeholders, whereby one is enabled to learn and apply various interviewing techniques and to listen actively.

BODYSTORMING rules stimulate the acceptance of rules in order to have a constructive, fair and creative working atmosphere. Even more so, graphical abilities are fostered by drawing ideas in accordance with the principle of visualization since images transport a meaning more precisely and faster. Clustering and selecting ideas unfold individual and team decision-making processes.

Prototyping an idea trains putting thoughts into action and learning how to communicate ideas. Using different forms of prototypes benefits haptic logic while building it, and opens the horizon to deliver an idea in distinct ways. This approach of several trials allows failures and deals with them playfully Further, it enables integrating and implementing user feedback from the TESTING to ITERATION in general.

Miscellaneous presentation tools are discovered and tried out as well as presentation skills being developed by the self-experience of presenting, but also by seeing the presentation of others. These team interactions in general thus activate how to cope with pushing forward one's own ideas and how to generally behave under certain social circumstances as well as under pressure (assertiveness, ability to decide on a behavior, ability to behave flexibly, self control).

During the whole design thinking process the ability to organize oneself as a person and in a team is practiced and improved through the freedom of guided self-regulation.

4 Do Our Theories Prove to be Resilient in Reality?

Having collected these theoretical frameworks and having gained many insights from interviews held at schools, ministries and with students, we made up numerous hypotheses we wanted to check in a real-life environment. In the following, we want to give you a short introduction into our experimental setup and then present five of our insights. For further information please consult the dissertation by Christine Noweski (Noweski 2012).

In order to observe realistic school settings, we decided to bring typical design thinking work style into regular schools.

The experiment therefore took place at a public *Gymnasium* in Potsdam with the full support of the principle and teachers of level 10. Level 10 is the last level, all students in Germany take together, before deciding to go on with Abitur, in preparation for a university admission, or to continue with a professional training.

Fig. 7 Design Thinking workspaces in the classroom (Photographer: Fabian Schülbe 2011)

It was comprised of students who were 15 or 16 (though we had one student aged 14 at the beginning of the experiment).

We split up the whole level (4 classes and 116 students) into teams of 4 and 5 students and had them work for 3 days on the challenge: *What and how can teachers profit from students knowledge as digital natives?* in a typical, flexible design thinking space, also used by the *School of Design Thinking* in Potsdam. The workspaces, consisting of two moveable whiteboards, a moveable high table and two highchairs (for up to five team member, so standing most of the time was inevitable) were brought into regular classrooms (Fig. 7).

Twelve of the 24 teams were supported by six teachers in training following Dewey's instructions, and 12 supported by six design thinking coaches. All coaches (Dewey and design thinking instructions) were chosen on the basis of having no particular knowledge in the subject of the workshop (digital media), being young (between 24 and 28) and motivated. The coaches were prepared in a training session. Here, they got information to intensify their already existing knowledge on their pedagogical approach.

We told the students when they arrived the morning to which teams they had randomly been assigned (giving attention that gender and classes were dispersed as equally dispersed as possible). There was a facilitator for each room (six teams), supporting the teacher and students with organizational and methodological difficulties, but the main challenge was left to the coaches and students themselves. They knew their challenge, the time frame and the method they ought to use and all of them were told to have as much fun as possible.

Everyday, students and teacher had to fill out several questionnaires, but spending no more than 20 min altogether per day on it, except for the "Social Competencies Inventory" (ISK 2009, see chapter "The Faith-Factor in Design Thinking: Creative Confidence Through d.school Education?" *How does Design*

Question answered by teachers: How did the students come across throughout the workshop?

	-3	-2	-1	0	1	2	3	
	☐	☐	☐	☐	☐	☐	☐	
more interested than normally at school								less interested than normally at school
more receptive than normally at school								less receptive than normally at school
less independent than normally at school								more independent than normally at school
more friendly than normally at school								less friendly than normally at school
less engaged than normally at school								more engaged than normally at school
less emotionally involved than normally at school								more emotionally involved than normally at school

Fig. 8 Average teacher judgments regarding the question: "How did the students came across throughout the workshop?" rated on a scale ranging from − 3 to + 3; negative values indicate the *left* characterization applies more; positive values indicate the *right* characterization is more applicable

Thinking contribute in developing twenty-first century skills?), which was filled out by the students in their regular class settings before and after the workshop.

To see what impact the workshop had – if any – on the social skills of students, pre-post comparisons (that is: gain-scores) were calculated. In summary, students of the design thinking condition profit more than students of the Dewey-condition. Even though not all differences in gain-scores are large enough to reach statistical significance, the picture is pretty consistent: In an 18 out of 21 scale the gain-scores are more favorable for design thinkers. In particular, the gain-scores differ with statistical significance ($p < .05$) on the following scales, favoring design thinking: Self-Expression, Direct Self-Attention, Self-Monitoring and Reflexibility. Close to significant ($p < .1$) are differences of gain-scores on the following scales: Assertiveness, Flexibility of Action, Indirect Self Attention and Person Perception.

1. Teachers describe the students as more participatory than usual at school if a constructivist teaching method is applied (see Fig. 8).
2. Teachers consider Design Thinking a highly valuable teaching method– more valuable than the Dewey approach (see Fig. 9).
3. Teachers state they are very likely to pursue a Design Thinking project if possible. Whether they would carry out a Dewey project is much less certain (see Fig. 10).
4. The teacher-student relation is positive in Design Thinking and in Dewey projects. In Design Thinking projects it is even more positive than in Dewey projects, and this consistently so (see Fig. 11).

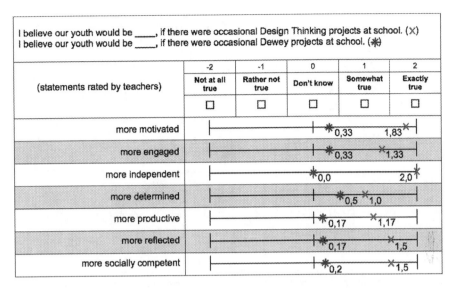

Fig. 9 Average teacher judgments regarding the expected impact of Design Thinking or Dewey's project work at school

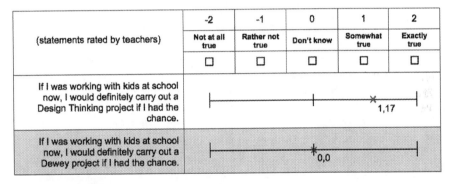

Fig. 10 Average teacher statements regarding whether or not they are likely to carry out a Design Thinking or Dewey project at school

5. Students appreciate the Design Thinking and the Dewey method. Consistently, they value the Design Thinking method even more than the Dewey method (see Fig. 12).
6. Mood assessment (see Fig. 13)

   On each workshop day, students and coaches specify their mood: in the morning, at midday and in the afternoon. The mood scale ranges from −10 (extremely negative) to +10 (extremely positive). There is one additional point of measurement for coaches due to their day of preparation ahead of the workshop.

   Students and coaches report positive sentiments throughout the whole project. Indeed, at each single point of measurement all four groups (students Dewey,

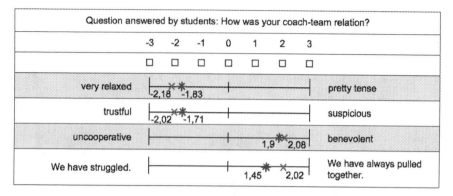

Fig. 11 Average student ratings of coach-team relation in Design Thinking (×) versus Dewey (∗) projects

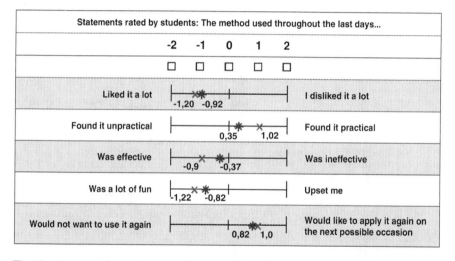

Fig. 12 Average student ratings regarding the Design Thinking (×) versus Dewey (∗) method

students Design Thinking, coaches Dewey, coaches Design Thinking) report an average mood in the positive realm (above zero).

Daily trends. On all three project days there is a trend that the mood improves from morning to afternoon.

Final sentiments. Students leave the workshop with a very good sentiment both in the Dewey and in the Design Thinking condition. For the coaches, an immense difference becomes apparent: The mood of Dewey coaches drops drastically while that of Design Thinking coaches takes off.

All in all, we can conclude our hypotheses confirmed that a teacher would be more likely to repeat a constructivist teaching method in a real school scenario

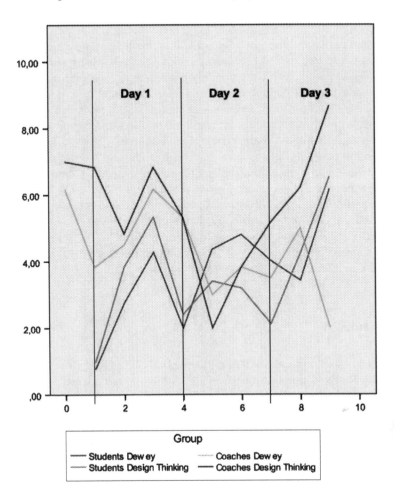

Fig. 13 Positive sentiments

when applying the design thinking process. And not only that, but surprisingly for us, the students of the design thinking condition profited more than the students of the Dewey-condition. So, the impact of Design Thinking in teaching in schools is even stronger than we expected. Students and teachers profit from it and the Department of Education's requirements (as demanded by society and economy) are being fulfilled.

5 Where Do We Go from Here?

Theoretical findings about the advantages and the use of constructivist learning and criteria for its realization are clear (Reich 2008; Dewey 1916). The practical implementation itself, however, is not yet being implemented effectively (Gardner

2010; Wagner 2011). Teachers seem to be demotivated and helpless in realizing holistic project work, and using constructivist methods, partly because of the absence of feedback, partly because of difficulties in assessing performance, as well as a lack of recommendations of designing constructive learning, according to the individual needs of their classes. We therefore conclude: there is a missing link between theoretical findings and demands and practical implementation of constructivist learning and teachings. This has led teachers to focus on approved and easily assessable content learning methods, and mostly deny affective, moral attitudes and practical, instrumental skills (Himmelmann 2005, also see Fig. 1) which however are a crucial fundament of the development of twenty-first century skills. Wagner refers to this as the *Global Achievement Gap*, the gap between *what even the best schools are teaching and testing versus the skills all students will need for careers, college, and citizenship in the twenty-first century* (Wagner 2011). We claim that, Design Thinking as constructivist methodology offers teachers the needed support towards a new way of teaching. Through a formalized process it may serve as a bridge between demand and reality.

6 Thanks

We sincerely want to thank the Hasso Plattner Design Thinking Research Program, especially Hasso Plattner, who believes in the value of our research, the board for funding us for 3 years and our dear colleagues for the fruitful discussions and insights.

References

Amabile TM, Khaire M (2008) Creativity and the role of the leader. Harv Bus Rev 86(10):100–109
Ananiadou K, Claro M (2009) 21st century skills and competences for new millennium learners in OECD countries, OECD education working papers, No. 41, OECD publishing. Available at: http://dx.doi.org/10.1787/218525261154. Accessed 11 Nov 2011
Brown T (2006) Presentation on innovation through design thinking at MIT World Forum, 16 March 2006. Available at: http://mitworld.mit.edu/video/357/. Accessed 11 Nov 2011
Brown T (2008) Design thinking. Harv Bus Rev 86(6):84–92
Buchanan R (1992) Wicked problems in design thinking, design issues, The MIT Press, 8(2):5–21. Available at: http://www.thestudiony.com/ed/bfa/Rittel+Webber+Dilemmas.pdf. Accessed 11 Nov 2011
Carroll M, Goldman S, Britos L, Koh J, Royalty A, Hornstein M (2010) Destination, imagination and the fires within: design thinking in a middle school classroom. Int J Art Design Educ 29(1):37–53
Cedefop (2008) Terminology of European education and training policy. A selection of 100 key terms. Luxembourg: office for official publications of the European communities. Available at: http://partners.becta.org.uk/upload-dir/downloads/page_documents/research/ict_attainment04.pdf. Accessed 11 Nov 2011
Cisco/Intel/Microsoft assessment and teaching of 21st century skills project. www.atc21s.org. Accessed 11 Nov 2011

Dewey J (1916) Democracy and education: an introduction to the philosophy of education. MacMillan Company, New York

Dewey J (1931) Ausweg aus dem pädagogischen Wirrwarr. Inglis Vorlseung 1931. In: Petersen P (ed) (1935) Der Projekt-Plan. Grundlegung und Praxis von, John Dewey und William Heard Kilpatrick, Weimar, pp 85–101

Drucker PF (1985) The discipline of innovation. Harv Bus Rev 1985:67–72

Edelman J (2011) Understanding radical breaks: media and behavior in small teams engaged in redesign scenarios. Ph.D thesis, Stanford University (to appear in December 2011)

Erdmann J (2010) Design thinking und multisensuelles Lernen Praxisbeispiele zur Gewaltprävention und zum sozialen Lernen, Schriftliche Hausarbeit im Rahmen der Ersten Staatsprüfung für das Lehramt für die Bildungsgänge der Sekundarstufe I und der Primarstufe an allgemein bildenden Schulen, Landesinstitut für Lehrerbildung, Berlin

Fagerberg J (2003) Innovation – a guide to the literature. In: Working papers on innovation studies, Centre for Technology, Innovation and Culture, University of Oslo. Available at: http://ideas. repec.org/s/tik/inowpp.html. Accessed 11 Nov 2011

Freeman C, Soete L (1997) The economics of industrial innovation, 3rd edn. Routledge, London/ New York

Gardner H (2007) Five minds for the future. McGraw-Hill Professional, New York

Hasselhorn M, Gold A (2009) Pädagogische Psychologie: Erfolgreiches Lernen und Lehren. Kohlhammer, Stuttgar

Himmelmann G (2005) Was ist Demokratiekompetenz? Ein Vergleich von Kompetenzmodellen unter Berücksichtigung internationaler Ansätze, in Edelstein, Wolfgang und Fauser, Peter: Beiträge zur Demokratiepädagogik, Schriftenreihe des BLK-Programms "Demokratie lernen & leben", Editor Himmelmann, Gerhard: Berlin. Available at: http://blk-demokratie.de/materialien/ beitraege-zur-demokratiepaedagogik/himmelmann-gerhard-2005-was-ist-demokratiekompetenz. html. Accessed 11 Nov 2011

IDEO (2011) The design thinking toolkit for educators. Available at: http://designthinkingfore ducators.com/#toolkit. Accessed 11 Nov 2011

Kanning UP (2009) ISK – Inventar sozialer Kompetenzen, Manual and test. Hogrefe, Göttingen

Kilpatrick WH (1918) The project method. Teach Coll Rec 19:319–323

Knoll M (1991) Lernen durch praktisches Problemlösen. Die Projektmethode in den U.S.A., 1860–1915. In: Zeitschrift für internationale erziehungs- und sozialwissenschaftliche Forschung 8:103–127

Kolb DA (1984) Experiential learning: experience as the source of learning and development. Prentice-Hall Inc., New Jersey

Miller A (2010) Realism. In: Edward Z N (ed) The Stanford Encyclopedia of Philosophy, Summer 2010 edn. Available at: http://plato.stanford.edu/archives/sum2010/entries/realism/. Accessed 11 Nov 2011

Noweski C (to appear in 2012): Mediation of democracy competencies through 21st century skill support at secondary schools (working title)

Open Ideo http://www.openideo.com (for the brainstorming rules: http://www.openideo.com/ fieldnotes/openideo-team-notes/seven-tips-on-better-brainstorming). Accessed 23 Nov 2011

Partnership for 21st skills www.21stcenturyskills.org. Accessed 11 Nov 2011

Paulus PB (2000) Groups, teams, and creativity: the creative potential of idea-generating groups. Appl Psychol: Int Rev 49(2):237–263. Available at: http://web.ebscohost.com/ehost/detail? vid=1&hid=17&sid=21ba8342-ea0e-465a-b4b1-02190525481c@SRCSM2&bdata=JnNpd GU9ZWhvc3QtbGl2ZQ==#db=pbh&AN=3263821. Accessed 11 Nov 2008

Peirce CS (1901) On the logic of drawing history from ancient documents especially from testimonies, collected papers, vol. 7

Pink DH (2006) A whole new mind: why right-brainers will rule the future. Penguin Group, New York

Reich K (2008) Konstruktivistische Didaktik: Lehr- und Studienbuch. Beltz, Weinheim

Rittel HWJ, Webber MM (1973) Dilemmas in a general theory of planning, vol 4, Policy sciences journal. Elsevier Scientific Publishing Company, Amsterdam, pp 155–169

Runco MA (2004) Creativity. Annu Rev Psychol 55:657–687

Rychen DS, Hersch SL (eds) (2003) Key competencies for a successful life and a well-functioning society. Hogrefe & Huber, Cambridge, MA

Schumpeter JA (1961) The fundamental phenomenon of economic development, in: the theory of economic development. Oxford University Press, New York, pp 57–94

Wagner T (2010) The global achievement gap: why even our best schools don't teach the new survival skills our children need–and what we can do about it. Basic Books, New York

Weinert FE (2003) Concept of competence, OECD 1999 (not citeable). Definition und Auswahl von Schlüsselkompetenzen, Zusammenfassung PISA Bericht, OECD 2003

"I Use It Every Day": Pathways to Adaptive Innovation After Graduate Study in Design Thinking

Adam Royalty, Lindsay Oishi, and Bernard Roth

Abstract As the demand for creative and adaptive workers grows, universities strive to develop curricula that enable innovation. A pedagogical approach from the field of engineering and design, often called design thinking, is widely thought to foster creative ability; however, there is little research on how graduates of design thinking programs develop and demonstrate creative skills or dispositions. This chapter proposes a new model for the development of creative competence through design thinking education, and investigates alumni outcomes from a graduate school of design thinking. Quantitative and qualitative data from a survey (N = 175) and in-depth follow-up interviews (N = 16), indicate that alumni apply a range of design thinking methods and dispositions in their professional lives, particularly related to creative confidence, comfort with risk and failure, and building creative environments. We explore potential mechanisms by which students develop these capacities and foreshadow future analysis of obstacles to innovation in the workplace.

1 Introduction

Society faces increasingly complex challenges that require more than the traditional set of bounded skills. With innovation and a "creative class" of professionals at the forefront of economic and social progress (Florida 2002), instilling the disposition and ability to adapt *is* the future of learning (Krumboltz 2009). With higher education under pressure to prove its value (e.g., Barreca 2010), institutions and researchers are responding by placing a greater focus on one of this conference's central questions: "How do we teach, not only for factual memory and procedural

A. Royalty (✉) • L. Oishi • B. Roth
Stanford University, California, USA

H. Plattner et al. (eds.), *Design Thinking Research*, Understanding Innovation,
DOI 10.1007/978-3-642-31991-4_6, © Springer-Verlag Berlin Heidelberg 2012

skills, but also for adaptive and flexible understanding that can be used beyond formal schooling and throughout life?"

An answer to this question that has become popular over the past 20 years comes from the field of engineering and design, and is known as "design thinking". Although derived from both engineering and the social sciences, design thinking has developed into its own domain, and has become influential in a wide array of disciplines such as business, health care, and education (Nussbaum 2009). More than 30 institutes in 15 countries have added design or design thinking to their curriculum in the past 5 years, with many including outreach to K-12 education.

While there is not a single comprehensive definition of design thinking, it is known as a human-centered problem-solving process. Brown (2008) identifies three stages in design thinking: inspiration, ideation and implementation. Inspiration includes understanding the problem, doing research, and organizing information synthetically; ideation refers to brainstorming, prototyping, testing, and re-designing; finally, implementation is execution of the solution, with the recognition that implementation leads to new projects or the next iteration of the current one. Characteristics of design thinkers, according to Brown, include empathy (taking others' perspectives), integrative rather than solely analytical thinking, optimism, experimentalism, and seeking collaboration.

Although much work has been done on design education, particularly as it relates to the arts or to engineering (Sheppard et al. 2008), the success of design thinking for teaching a process-based ability to innovate, regardless of discipline, has not been evaluated empirically. One of the biggest challenges of evaluation is defining and measuring the desired outcomes. Nussbaum (2011) writes, "Design Thinking was a scaffolding for the real deliverable: creativity," and argues for a move to what he calls creative intelligence, or "the ability to frame problems in new ways and to make original solutions."

In this study, we define the goals of design thinking education as developing creative intelligence or competence, including skills, confidence and performance in relation to real-world problems. To begin investigating how these goals may be achieved, we surveyed and interviewed alumni of a well-established institute of design and design thinking. Of particular interest were graduates' applications of design thinking methods in their work, the personal and professional results of their applications and adaptations of these methods, and how their educational experiences may have contributed to their abilities as productive creative professionals.

2 Theoretical Framework

Needless to say, the literature on creativity is vast, and there is not sufficient space in this chapter to review it all. Furthermore, we are not attempting to define (or re-define) creativity; rather, the research takes as foundational the idea that design thinking *is* a creative process with the potential to produce creatively-abled

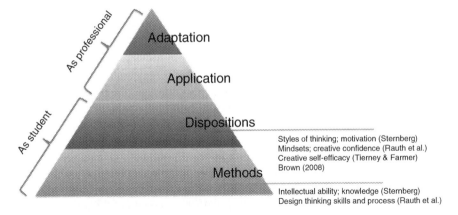

Fig. 1 Conceptual model for stages of achievement in design thinking

professionals. Therefore, the literature we examine has more to do with the *learning process* of design thinking and creativity, rather than the nature of these constructs.

We began with Sternberg's (2006) identification of six prerequisites for creativity: intellectual abilities, knowledge, styles of thinking, personality, motivation, and environment. We also wanted to consider Tierney and Farmer's (2002) concept of creative self-efficacy, which refers to an individual's belief in his or her own capacity to be creative. Finally, we considered Rauth et al. (2010) proposal of design thinking as a pathway to creative confidence. In their model, design thinking methods such as empathy, brainstorming, and prototyping are the basis of a pyramid of skills, with creative confidence at the very top. The second and third levels, after basic methods, are process fluency and more general design-oriented mindsets, such as being "human-centered" and having a "bias towards action".

Synthesizing the psychology and design sides of the previous research on the development of creative capacity, we propose a new model of learning through design thinking (Fig. 1). In this model, the levels of the pyramid represent engagement in design thinking over time, with the bottom being basic or superficial, and the top indicating deep or advanced understanding and usage. The first level, methods, refers to design thinking skills and processes, and draws on both Sternberg's and Rauth et al.'s frameworks. The second level, dispositions, include Rauth et al.'s construct of creative confidence (or Tierney and Farmer's creative self-efficacy) as well as other more holistic attitudes, and Brown's (2008) experimentalism (also known as "risk-taking"). By practicing the methods and developing the dispositions of design thinking, individuals can move into the third level, applying design thinking creatively towards solving real problems in life and work. The final stage is adaptation: going beyond what was learned and using design thinking knowledge and capacity in novel and unexpected ways. These last two stages are explicitly focused on successful real-world applications of design thinking *after* schooling, particularly in professional endeavors. This conceptual framework directly influenced our decision to focus on alumni rather than current students, as well as using them as the lens through which we approached the data.

3　Methods

3.1　Design

The research took place at the Hasso Plattner Institute of Design at Stanford University (known as the "d.school"), which has offered over 60 graduate-level courses since 2006. Though diverse, these courses all involve inter-disciplinary teams applying design thinking to real-world challenges across diverse fields, and are open to graduate students in all disciplines.

This study used a mixed methods approach, including an online questionnaire and follow-up qualitative interviews. The research was both exploratory and confirmatory: a priori hypotheses about the specific design thinking skills and mindsets taught at the school were tested, but open-ended questions also probed for unexpected outcomes.

3.2　Participants and Procedure

A survey invitation was emailed to 670 d.school alumni; the response rate was 28 % (N = 184). Of those who reported gender, 73 (56 %) were men and 57 were women. Participants took the survey via a website (Qualtrics 2011) at their convenience. Respondents had finished graduate programs in business, engineering, arts and sciences, law and medicine between 2005 and 2011; the majority (83 %) were graduates from 2008 and after.

Based on responses to the questionnaire, 27 participants were invited for follow-up interviews. They were selected to represent a range of majors and current occupations. Interviewees were offered $50 gift cards to participate in the 1-h interview. A total of 16 participants (10 women) agreed to be interviewed either in person or online via Skype. Seven currently worked in business (including healthcare, technology and entertainment), three were self-employed or entrepreneurs, three worked in education, two in consulting or research, and one was in engineering.

4　Materials

The survey primarily investigated perceived outcomes and applications of d.school education in professional life. Two key questions were, "In the last month, how often did you apply what you learned at the d.school at work?" (1-"Not in the Last Month" to 5-"Almost Every Day") and, "To what extent do you think your d.school experience makes you different from peers without this experience?" (1-"Not at all" to 5-"A great deal"). In addition, the survey asked participants for examples of how

Table 1 Interview topics and main questions

Interview topics	Main corresponding question
d.school influences career/life path	Take just a minute to sketch out the journey that led you to the d.school, your time there, and what you did in your life afterwards. After you sketch this out, I'll ask you to share it with me
d.school influences on work life	Can you tell me about a time when you applied what you learned at the d.school at work?
d.school influences on personal life	Can you tell me about a specific time when what you learned at the d.school affected your personal life?
Comparisons with colleagues	Next, I'd like you to think about someone in your life who has not taken classes at the d.school. What are some of the ways that you see yourself as different from this person?
Salient d.school memories	What are your most memorable experiences from your time at the d.school?

they applied what they learned at the d.school at work, as well as explanations of the ways in which participants felt different from their colleagues.

The interviews were semi-structured to cover a few major themes: experiences at the d.school (which could vary from one course to being a paid teaching assistant), how interviewees used what they had learned at the d.school in their current occupation and personal lives, and their most important memories and learning experiences from their courses (Table 1). There were two interviewers who completed 11 and 5 interviews respectively, after cooperatively developing the protocol and discussing interview methods. Two of the interviews were considered pilots; after each interview, the researchers discussed the questions and responses in order to refine the protocol. Both interviewers also coded all of the interview data individually. The coding categories included instances of application and adaptation and the most salient three to five d.school attributions each interviewee made throughout the interview. Disagreements were discussed until consensus was reached (grounded convergent coding).

5 Results

5.1 Survey

Survey respondents said that they used what they had learned at the d.school, on the average, once to 2–3 times per week ($M = 2.52$ out of 5, $SD = 1.37$). They also rated their difference from peers as 2.66 ($SD = 0.97$) out of 5, which semantically is closest to "A lot".

In the open-ended questions, the most frequently reported methods were empathy (n = 56, 43 %), prototyping (n = 49, 38 %) and brainstorming or ideating (n = 48, 37 %). The final two categories, teamwork/collaboration and defining/re-framing problems, were less frequently mentioned (by 19 % and 14 %, respectively).

Table 2 Examples of design thinking methods and dispositions (reported by survey participants)

Design thinking skill	Example from data
Defining/re-framing	"There are more possibilities, and you see more opportunities in what you had become jaded towards"
Brainstorming/Ideating	"A colleague and I were trying to create a report, and we first made a giant list of all of the possibilities, and then voted"
Empathy	"Willingness to devote time to gathering customer insights"
Prototyping	"I didn't wait to perfect the service before launching it, I just launched it as soon as I could, solving the kinks as they come"
Teamwork/Collaboration	"Caring less about being personally correct or incorrect and instead focusing on coming to a working solution as a team"
Creative confidence	"I have a different approach to innovation: innovation is inside everyone, our job is to create the right prototypes to grab [it]"
Bias towards action	"I have a willingness to experiment and say yes, let's try it!"
Comfort with uncertainty/ failure	"I try to fail early and often in order to succeed later"

Table 2 gives examples of each category. In addition to specific methods responses were also coded for three key dispositions, identified a priori. The most frequently referenced disposition was creative confidence, and was brought up by 25 % of participants. A similar number (24 %) talked about becoming comfortable with risk, ambiguity, change or failure. Having a bias towards action (in contrast to, thinking or planning) was described by 11 % of respondents.

A theme that emerged during coding was the participant's role as a teacher or leader when using design thinking in their work. Of the 130 respondents who completed the open-ended questions, 34 (26 %) described sharing concepts, processes, skills and mindsets learned at the d.school with others in a professional context.

6 Interviews

6.1 Dispositions

Given that the survey demonstrated a variety and depth of methods and skills, we began analysis of the interviews at the next level - dispositions. The two dispositions reported most frequently by interviewees were creative confidence and comfort with risk or failure.

All participants attributed some level of confidence in their creative abilities to their time at the d.school. This came up as an explicit outcome; as one participant put it, "Thinking different [sic], being more creative, exchanging ideas with others to get feedback and shape my thoughts." Creative confidence, as evidenced by creative action, also came up when alumni compared themselves to colleagues:

"When [friends] see a problem, like, for example, if we're driving in the car together and there's terrible traffic. They're a lot more willing to sit back and say there's terrible traffic...It seems like sometimes they're more willing to accept the situation they're given, whereas if I'm in that situation I'll try to find a way out." Finally, alumni attributed creative actions they took after graduation to their d.school experience. One participant noted that after she finished at the d. school, she used the idea of "failing early" when creating and releasing her first cooking DVD.

The example of the cooking DVD also shows the second disposition attributed to the d.school, which appeared in nearly all the interviews: comfort with risk and its consequences, including failure. Risks came up in two general categories: large career or life changes, and smaller risk strategies in daily professional and personal activities. One alumna, who changed careers from engineering to education, said that a big outcome of her time at the d.school was a new courage to experiment. As she put it, "The other way [I changed] is like, not feeling afraid to try new things, and not feeling like, 'Oh, I'm gonna try teaching, and if it's not the right choice for me it's the end of the world,' and just iterating on my personal life and iterating on my professional life." An example of comfort with risk on a smaller scale came from an alumnus who worked on a website, who said, "There's just a lot of garbage on the site that we stopped supporting 2 weeks after we released it 'cause oh, it failed. Okay, don't worry about it. Let's just go on to something new." In his view, failure was an acceptable part of the development process.

6.2 Application and Adaptation

For this study, we defined applying design thinking as using methods and dispositions learned at the d.school in another context without noticeable modification. For example, many participants mentioned leading brainstorming sessions, which they learned at the d.school, in their new work teams or in other groups after graduation. One alumna noted, "[my colleagues] love to be taught how to brainstorm. And when you kind of introduce some rules of brainstorming, they're really grateful."

Adaptation, on the other hand, was defined as significantly modifying skills, procedures, and processes taught at the d.school, and then using them in a new context. One alumnus spoke about modifying brainstorming into multiple forms, beyond what was taught at the d.school. When asked for an example of how he used design thinking, he said, "[I] set the rules for brainstorming [at team meetings]. So there are different ways of brainstorming...I've experimented with them many times."

In each of the 16 interviews, alumni detailed applying or adapting a variety of design thinking methods in their lives. Examples include brainstorming with other employees, testing prototypes with real users, and choosing to share information in a visual manner rather than in the companies' standard presentation formats. The different types are based on the design thinking skills laid out in Table 2. Every interviewee demonstrated using a design thinking activity that fit in to at least four categories. Some subjects recalled activities that fit in all eight categories. A pattern that existed uniformly across participants is that applications appeared much more often than adaptations.

A commonly reported application of design thinking was getting feedback on unfinished products or ideas from other people, which encompass methods of prototyping and iteration. A quotation from an alumnus illustrates this point. He said, '[the d.school] helped me by allowing me to do something without being perfect. This was not perfect at all, but I did it. So just doing something, prototype – that was a prototype... And prototypes change, and that's okay.'

One of the key adaptations of design thinking that alumni reported was developing Creative Environments (CE) in their work or personal lives. Fostering a creative environment frequently took the form of teaching coworkers elements of design. "And afterwards I graduated [sic], I was well-versed enough to start teaching the [design thinking] method again to other folks." Fusing parts of the design thinking process within an existing work structure was also considered building a creative environment. One interviewee noted, "Because I've been in the company longer than a lot of people here I tend to be an influencer, so I basically went to where I had a product at that time and said, 'Let's do these interviews.'" As this participant also noted that doing user research, such as interviews, was a key learning point from his time at the d.school, this example clearly showed that he was transferring his educational experience to his professional endeavors. Another example was creating actual spaces that support creative work. "I purposely have a studio apartment that has a lot of colors... and I try to have areas where I can go and read something and not look at my computer at all. And there's areas where I can draw and there's other areas where I can rest, which I got out of the [d.school] trailers." For this alumna, allowing herself to be creative at home meant consciously designing the space to support this goal.

7 Potential Mechanisms

To investigate how alumni gained the skills and dispositions that resulted in their remarkable abilities to apply and adapt design thinking after they left the d.school, we focused on their explanations of their development and particularly, their most salient or impactful memories. The most commonly mentioned themes that alumni spontaneously brought up were: working on *real applied problems*, doing *field work* (interviewing and testing ideas with people they did not know), their *project demonstrations* (presenting their work to their classes), and working on *cross-disciplinary* teams. Two other types of memories that came up often, but less frequently than the first four, regarded *inspirational people*, often guest speakers from industry or corporate sponsors (e.g., Walt Disney), the general *energy* or culture of the d.school space.

Just as interesting as what alumni remember is what they do not remember (or spontaneously report). They rarely mention specific lectures on design thinking content nor do they especially recall domain-specific lectures, which happen throughout classes that combine design thinking with topics like education and environmental responsibility. Finally, there was little to no talk of grading or other interactions (such as receiving feedback or guidance) with instructors.

8 Discussion

Though we unfortunately did not have space in this chapter to fully describe the d.school and its curriculum, nor report all of our findings, we have presented the beginning of a research program that will offer valuable perspectives to educators interested in making post-secondary education more conducive to graduates' ability to innovate at work and in society. Specifically, we have shown several key learning outcomes of a design thinking program, which have persisted for several years after graduation and have been implemented in graduates' workplaces.

From the alumni survey, we learned that the most frequent skill- or methods-based outcomes of the d.school were empathy for users/clients, brainstorming/ideating, and prototyping/iterating; alumni also reported applying what they learned about inter-disciplinary teamwork in their professional lives. It is also interesting to note that alumni entered diverse fields such as social entrepreneurship, medicine and law – few became product designers – and most reported propagating the dispositions found in the d.school in more conventional work environments such as large corporations, research universities, and secondary schools.

Looking at the results from the interviews regarding alumni dispositions, we found themes of creative confidence and comfort with risk and failure. Linking creative confidence to the most commonly reported memories from the d.school, we believe that one way students build confidence in their own ability to be creative and act creatively is through going through the project cycle associated with the real-world problems given in d.school classes. Another explanation is social: the inspiring people that students remembered encountering, along with the intangible energy of the d.school space and culture, may free or enable people to act in creative ways. This explanation is confluent with Sternberg's (2006) argument that creativity is, essentially, a *decision* to be creative, which requires social sanctioning. Finally, alumni spoke about the powerful effect of presenting their projects to classmates at the end of each academic period. It is possible that the act of viewing one's own creative output, and receiving positive feedback from colleagues, builds creative confidence.

Based on our findings, we are developing a creative confidence self-efficacy measure and testing it with current d.school students. Items were created based on categories that arose from the survey data. They include the confidence to develop a creative environment, comfort with ambiguity, and how actively one is continuing to develop a creative process. The goal is to create a reliable measure that shows growth in creative confidence over the course of a d.school learning experience, such as a class or workshop. This would greatly enhance our ability to experiment with one of these types of experiences in an effort to determine what exactly affects students' creative confidence.

To understand the second disposition theme, how alumni became comfortable with risk and failure, we also considered their most frequently reported memories from the d.school, along with the d.school curriculum and pedagogy. As many participants spoke about the field work they were required to do (e.g. interviewing strangers about their product prototypes), this is one potential path to getting over fears of risk or failure. Another aspect of the d.school curriculum, mentioned by many interviewees, was the process of getting feedback from others. This is personally risky because the results are unknown and students often feel a sense of vulnerability during these exercises. But throughout the projects, students become accustomed to this by lowering the stakes of each interaction while increasing the total number of interactions they have.

One of the most interesting outcomes of the alumni interviews was finding how graduates adapt design thinking in their personal and professional lives. We felt that their tendency to build creative environments was an important outcome because it shows how creativity in an educational context may be transferable to the professional context. Alumni were willing to put considerable effort into supporting their own creative capacity, including teaching design thinking to others, working for new processes at their workplace, and (when necessary) changing occupations.

A question we would like to pursue through further analyses of the interview data, as well as new research on alumni outcomes, concerns the challenges many alumni brought up regarding being creative or fostering design thinking in their workplace. Though d.school students learn to work on "real-world" problems, often in concert with an industry sponsor, the post-graduation working context is clearly different. Furthermore, the constraints *students* experience are very different from those faced by engineers, consultants, designers and other employees of large corporations or institutions with defined (and often demanding) clients. Several alumni described resistance to processes they learned at the d.school. One said, "It was slightly more acceptable to ask 'why' in the context of the d.school, because people kind of expect that kind of a question, that questioning, but when you move outside of the d.school, you can't just ask why." Future research will explore this theme of resistance and barriers to innovation. One approach that we plan to take is to conduct a similar study of alumni of d.school executive education programs. After learning about design thinking, these mid-career professionals return to the workforce in a significantly higher position than the average university graduate. Understanding this population should further illuminate what hinders the application and adaptation of d.school principles.

As we continue to analyze our data and conduct new research on enabling creativity through education in design thinking, we hope it will inspire others to consider trying innovative curricula. In particular, we want to emphasize the powerful experiences shared by the d.school alumni we interviewed, as pedagogies that can be applied in other settings to encourage creative ability and performance.

The impact of real-world projects, fieldwork, and inter-disciplinary collaboration have been recognized before; this research extends this trend to how these educational methods drive innovation in a diverse range of professional fields. We also hope to stimulate other institutions and programs interested in design thinking, creativity, or innovation, to continue adding to our knowledge of what forms of education are successful not only in school, but also for graduates working in an increasingly demanding, challenging and complex world.

References

Barreca G (2010). Is there a doctorate in the house. The chronicle of higher education. Accessed online at http://chronicle.com/blogs/brainstorm/is-there-a-doctorate-in-the-house

Brown T (2008). Design thinking. Harvard business review. Accessed online at http://hbr.org/2008/06/design-thinking/ar/1

Florida R (2002) The rise of the creative class: and how it's transforming work, leisure, community and everyday life. Basic Books, New York

Krumboltz JD (2009) The happenstance learning theory. J Career Assessment 17:135–154

Nussbaum B (2009) Design thinking battle–managers embrace design thinking, designers reject it. Bloomberg Businessweek. Accessed online at http://www.businessweek.com/innovate/NussbaumOnDesign/archives/2009/07/design_thinking_3.html

Nussbaum B (2011) Design thinking is a failed experiment, so what's next? Accessed online at http://www.fastcodesign.com/1663558/

Rauth I, Koppen E, Jobst B, Meinel C (2010) Design thinking: an educational model towards creative confidence. In: Proceedings of the 1st international conference on design creativity

Sheppard SD, Macatangay K, Colby A, Sullivan WM (2008) Educating engineers: designing for the future of the field. Jossey-Bass, San Francisco

Sternberg RJ (2006) The nature of creativity. Creativity Res J 18:87–98

Tierney P, Farmer SM (2002) Creative self-efficacy: its potential antecedents and relationship to creative performance. Acad Manage J 45:1137–1148

Part II
Design Thinking Research in the Context of Distributed Teams

Tele-Board in Use: Applying a Digital Whiteboard System in Different Situations and Setups

Raja Gumienny, Lutz Gericke, Matthias Wenzel, and Christoph Meinel

Abstract Tele-Board is a digital whiteboard system that helps creative teams working together over geographical and temporal distances. The nature of Tele-Board's synchronized setup allows every connected partner from anywhere in the world to join in the action. Tele-Board is rooted in traditional metaphors, which are easy to implement and come naturally to the user. Additionally, it is possible to follow a common thread in the development of ideas from their inception to conclusion. With the History Browser, the path of creative development can be retraced, reiterated and resumed – from any point in time – a huge benefit in ordering work and reaching conclusions. In this article, we report on several situations and setups in which Tele-Board was used by different teams. We demonstrate how our software suite can be used with various hardware setups and show how well the tools work in practical application. Furthermore, we illustrate Tele-Board use by globally distributed student teams, in remote test settings, during a sustainability conference, and by teams who are primarily used to traditional whiteboards and pen and paper.

1 Creative Work in the Digital World

For global companies it is vital that they regularly come up with new, innovative ideas in order to ensure long-term competitiveness of the organization. With these economic goals in mind, another incentive would be solving inherent problems of this world. If an organization manages to develop products or services that truly

R. Gumienny • L. Gericke • M. Wenzel • C. Meinel (✉)
Hasso-Plattner-Institut (HPI), für Softwaresystemtechnik GmbH, Prof.-Dr.-Helmert-Str. 2-3, 14482 Potsdam, Germany
e-mail: raja.gumienny@hpi.uni-potsdam.de; lutz.gericke@hpi.uni-potsdam.de; matthias.wenzel@hpi.uni-potsdam.de; meinel@hpi.uni-potsdam.de

H. Plattner et al. (eds.), *Design Thinking Research*, Understanding Innovation, DOI 10.1007/978-3-642-31991-4_7, © Springer-Verlag Berlin Heidelberg 2012

Fig. 1 Working remotely with the Tele-Board system

support other organizations and improves individuals' lives, the probability of public and economic success grows.

While researchers write more and more books on the inner core of Design Thinking (e.g. Brown 2008; Cross 2007; Lockwood 2009; Martin 2009; Plattner et al. 2009), we focus on tools to support teams who are working in a way, as it is taught e.g. in the School of Design Thinking in Potsdam. While many factors constitute the success of Design Thinking, one of them is its methods' tangibility and easy to use tools that are understood worldwide: paper, pens, whiteboards and different material most of us know from kindergarten.

Problems arise, when continents and time zones separate the team that is working on a problem. With this in mind, we developed the Tele-Board system, which provides the possibility to work creatively over distances and still retains the feeling and working modes of traditional tools (Gumienny et al. 2011). People can work with whiteboards and sticky notes in a way they are used to and additionally have the advantage of digital functions that don't exist in the analog world. For remote settings, all whiteboard actions are synchronized automatically and are applied by every connected partner. Optionally, the teams may include a video conference between themselves and the distributed team: the translucent whiteboard is an overlay on top of the full screen video of the other team members. This setup lets everyone see what the others are doing and which content they are referencing. Additionally, gestures and facial expressions can be seen (see Fig. 1).

Most recently, we focused on testing and deploying Tele-Board in a variety of different situations and contexts. Besides conducting scientific experiments and usability tests, we also provided Tele-Board to teams who were collaborating over distances or made use of digital tools instead of traditional ones.

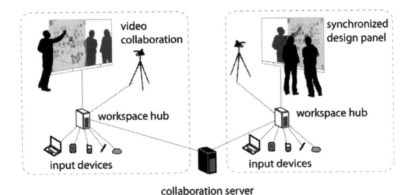

Fig. 2 General setup of the Tele-Board system

In this article, we will present Tele-Board's functions and possibilities with a special focus on different hardware setups depending on the given situation. Following this, we will report on experiences from different usage scenarios and their implications for our work. The next section will give an introduction to the design of the Tele-Board system.

2 Tele-Board: A Flexible System for Remote Collaboration

Tele-Board is a software system that supports remote collaboration based on electronic whiteboards. The interaction with the system works in a similar way to conventional whiteboards, i.e. writing, drawing, and erasing on the whiteboard surface can be done in the usual way. Beyond that, it is possible to create digital sticky notes using the whiteboard or additional input devices such as Tablet-PCs, iPads or smart-phones. At the whiteboard, it is possible to edit sticky notes, move, resize, and generate clusters of them.

Remote collaboration is facilitated by connecting several digital whiteboard devices at their corresponding locations with the help of the Tele-Board system as shown in Fig. 2. All of the actions mentioned above are synchronized automatically and propagated to every connected whiteboard. Every user can manipulate all sticky notes and drawings, no matter who created them. Furthermore, a videoconference feature is included. The whiteboard content can be displayed transparently on top of the full screen video of other team members. Local team members can see the actions and pointing gestures of the remote team members and vice versa, which facilitates an easier and more interactive session. The flexible architecture of the Tele-Board system makes it possible to start the whiteboard software on every computer (for more information see Gumienny et al. (2011)).

The content created with Tele-Board is organized based on *Projects*. A project can be used to embrace all phases of a design process. During the course of a

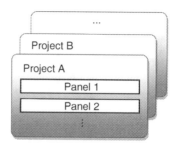

Fig. 3 Organization of Tele-Board content

traditional design project, a set of analog whiteboards is filled with sticky notes and handwriting. In Tele-Board, the digital counterpart of a physical whiteboard is called *Panel*. A panel is displayed with the help of an electronic whiteboard and can be filled with virtual content. The interrelation between panels and projects is depicted in Fig. 3. Every panel is assigned to a single project that in turn can contain an unlimited number of panels. Moreover, panels can be archived and restored to any state during the panel's progress (for more information see Gericke et al. 2010).

2.1 Tele-Board Components

The functionality of the Tele-Board software system is divided among different components, which are as follows:

Web Application: The web application[1] serves as an administration interface enabling users to maintain their projects and associated panels through a web browser. The whiteboard client that allows editing of a panel is started from this interface, what makes the web application the entry point of the Tele-Board system. Furthermore, it is convenient to use because there is no need to install any extra software.

Whiteboard Client: The Tele-Board whiteboard client is a platform-independent Java application. It facilitates whiteboard interaction, e.g. writing with different colors, erasing, and the creation of sticky notes. The client software runs on the user's computer, which can be connected to an electronic whiteboard. Therefore, it is possible to operate the system with any whiteboard hardware, a Tablet-PC or just with a mouse on a computer screen if no electronic whiteboard is available. Additionally, the whiteboard client interacts with the Tele-Board server component by synchronizing with other clients started at a remote location, as depicted in Fig. 4.

---

[1] http://tele-board.de/

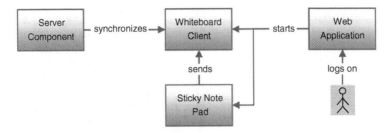

Fig. 4 Tele-Board components and their interrelation

Sticky Note Pad: This component can be used as a dedicated input tool as shown in Fig. 2. To increase flexibility in terms of input variety, we created different applications for writing sticky notes as an equivalent to paper-based sticky note pads. The Sticky Note Pad is well suited for Tablet PCs and other pen input devices. For convenient input from a handheld device you can use the corresponding App on an iPad (StickyPad HD[2]), iPhone or iPod touch (see Fig. 5).

Server Component: The server component coordinates all communication between the remote partners. All interactions are transferred as XMPP messages to keep the connected whiteboards synchronized. For advanced saving and resuming possibilities, we extended the server component with additional functions (see the History section for details).

2.2 Video and a Translucent Whiteboard Surface

Remote collaboration on electronic whiteboards can benefit from an accompanying videoconference showing the remote team interacting with their whiteboard. Without video, whiteboard interactions by remote team members appear as if made by a "ghost hand". For the current implementation, we decided to use Skype because of its proven reliability and ease of use. However, Tele-Board can be used with any third-party videoconferencing software. The whiteboard client can act as a translucent overlay on the video image giving the impression that the remote party is directly interacting with the whiteboard content (see Fig. 1). We also tried out different camera angles, for more information see Gumienny et al. (2011).

---

[2] http://itunes.apple.com/app/stickypad-hd/id464034808

Fig. 5 iPad running the StickyPad HD app as an input device for Tele-Board

2.3 Tele-Board History

As we learned from user feedback and interviews, people in remote teams very often work asynchronously. To support these working modes, we developed a solution helping team members, who cannot be connected at the same time, to understand what the others were doing and easily hand over their work.

Easy navigation through different whiteboard states and resuming work at any previous point in time is a major goal in our development process. A digital whiteboard solution can also offer the possibility of extensive and partly automated documentation. In traditional whiteboard settings it is time-consuming and troublesome to take detailed photographs after work is done. Written documentation for stakeholders and customers has to be prepared additionally. Another argument for the importance of implicit documentation is the statistical relevance for people researching team behavior and how design over distances and time differences is carried out. Various questions could be answered using the history data: What is the main working time of the employees? How can the output become measurable? Not only design researchers could be interested in this information, but also the designers themselves would profit from gaining insights into key factors of their creative work. The possibilities of the Tele-Board system in terms of the traceability of remote work concerning researchers are shown in Gericke et al. (2011).

The *Tele-Board History* is implemented as a central archive that is used to keep track of all data, make analysis more convenient, and enable asynchronous work. Therefore, communication data handled by the server is stored in a database. This allows the immediate analysis of the communication flow and storage of the real communication data rather than image representations of the content. For more information see Gericke et al. (2010).

Table 1 Variety of hardware in the observed setups

	Whiteboard hardware location one	Whiteboard hardware location two	Sticky note devices
Global student teams (Germany – California)	SMART board interactive display overlay	DELL interactive projector S300wi, rear projection	Laptops
Global student teams (Germany – France)	Luidia eBeam, Hitachi short-throw projector on the ground	DELL interactive projector S300wi, rear projection	Laptops
Logic grid puzzle	SMART interactive board UF45-680	SMART interactive board UF45-680	–
Sustainability congress	Panasonic elite Panaboard T8	Promethean ActivBoard	Three iPads, Laptop
Physical versus virtual boards	SMART interactive board UF45-680	Promethean ActivBoard	–
One day challenge with design thinking students	SMART interactive board UF45-680	SMART interactive display	Four iPads, TabletPC, Digital pen, Laptop

2.4 Flexibility Through Hardware Independence

In an ideal world, all teams, who want to work creatively over distances, would always have the best available hardware at hand: several state-of-the-art interactive whiteboards, different mobile devices for each user and an easy-to-use and reliable audio- and video conferencing system. Of course, in most working environments and situations this is not the case. The equipment is relatively expensive and often bulky, which can make logistics sometimes difficult.

Therefore, we always had hardware independence and flexibility in mind, when we designed Tele-Board. The web portal can be viewed with any browser, even on mobile devices. An installed Java Runtime Environment – as is the case on most computers – is the only requirement to use the whiteboard client. As all interactive whiteboard hardware and display technologies can emulate mouse input to the connected computer, it is possible to use any of them for working with Tele-Board. Of course, some devices behave more precisely or faster than others, but this often goes along with a higher price regarding costs and mobility.

For most situations in which we introduced Tele-Board, we used different hardware equipment – mostly because we had to adapt to the situation and equipment we found. In the following table, an overview of the various setups we used is presented (Table 1).

In the following, we will describe how the different hardware setups affected the teams' way of working and which devices are better suited for a particular situation.

3 Tele-Board in Use: Remote Location Setups

Our main objective during the early days of Tele-Board was the vision to support Design Thinking teams in distributed settings, mainly in the School of Design Thinking in Stanford and Potsdam. In recent years, Design Thinking and similar approaches have found widespread application. That is why we regularly get feedback from people looking for a solution such as Tele-Board. In order to develop an even more elaborate state of the system, we constantly try to use it in all kinds of setups and get feedback with regard to its usability – in addition to the scientific experiments we do for our research.

In the following two sections, we give an overview of what happened when we supplied very different teams with Tele-Board, while looking especially at their working modes and hardware setups. We start with the remote location setups.

3.1 Global Student Projects: The ME310 Course

ME310[3] is a course that has been taught at Stanford University for many years. Student teams work on real world design challenges from corporate partners. Teams of usually six to ten people are given 9 months of time, and the trust to design a complete package of innovation together with a corporate partner. This includes methods such as user observation, brainstorming, prototyping etc.

Over the years, ME310 has become more and more global. Most teams are distributed over the globe, e.g. joint teams of Hasso-Plattner-Institute students in Germany work together with their partner team at Stanford University or ParisTech University in France. Depending on their locations, the teams have very few opportunities for personal meetings together (usually only for kickoff and the final presentations). Meanwhile, these teams are separated across two countries.

In order to help them to communicate and collaborate over distances, we equipped the teams at HPI and ParisTech with interactive whiteboard hardware. In this first pilot year, there was no large budget for equipment and therefore we had to find low-cost setups. At the HPI in Potsdam, students used an interactive projector together with semi-transparent film in a rear projection setup (see Fig. 6, left). In Paris, we set up an eBeam system with a short-throw projector on the ground (see Fig. 6, right).

Unfortunately, the teams could not really work remotely as there were general problems concerning the Internet connection at ParisTech. So the students decided to have Skype calls at home.

At HPI, the students used Tele-Board for collecting and clustering their user research data, creating personas and brainstorming (see Fig. 7, left). They liked the

---

[3] http://me310.stanford.edu/

Fig. 6 Setups of the Tele-Board system for the global students project ME310 between Potsdam, Germany (*left*) and Paris, France (*right*)

possibility to easily add pictures and notes to the board and open it on any computer, not only at the office. However, they were missing a tool to easily draw sticky notes as our Sticky Pad App was still under development. They said that the "pen" of the interactive projector was too big and heavy for really using it on the whiteboard and therefore they did not write or draw a lot. In contrast, the Stanford team had a SMARTBoard device for one session and used it to make sketches and drawings, which show the large potential of using a well-fitting hardware (see Fig. 7, *right*).

3.2 Logic Grid Puzzle

Tele-Board can be used for synchronous as well as asynchronous work. The asynchronous aspect comes in two flavors: supporting designers to work in a team over different time zones, but secondly giving researchers a tool to analyze design teams by carrying out statistical analyses on the communication data. To point out the potential of those automatic analyses, we conducted a series of tests involving ten teams of two participants each. They used SMARTBoard devices in a synchronous setting with two locations – one user at each location. We were not so much focused on the results of the teams, but more on the process of analyzing the work of the participants and revealing the differences between the teams.

To make results more comparable, the task itself was quite structured and not as creative and design-focused as in other tests. We asked the participants to work on a logic grid puzzle solving a detective story, finding out who broke in at which house etc. As a starting point, a grid with the different attributes of each crime scene was already given, as well as a set of ten hints making it possible to fill out

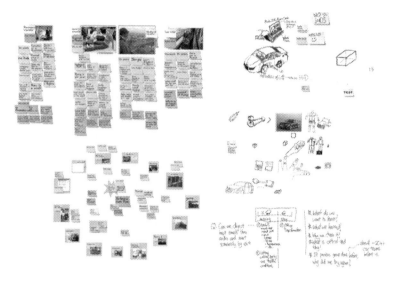

Fig. 7 Whiteboard content of the global students project ME310

the grid bit by bit. People were asked to collaborate on the solution, which means they often discussed certain steps of their solution process by referencing different cells of the grid, sorting the hints, and coordinating the completion of the solution.

We divided the teams into two groups: Those having a video-enabled connection (see Fig. 8), and those having video switched off and only relying on an audio connection. The hypotheses we wanted to address were as follows:

1. With a video-enabled connection, people are more efficient, i.e. solve the task faster.
2. With a video-enabled connection, people enjoy the task more, establish more common ground, more team spirit, and work together more closely.

We applied different analysis methods afterwards. Each of them has different possibilities and limitations (see Table 2). Our goal with the automatic analysis is not to replace traditional methods, but to enrich and save time on existing approaches by doing standard analysis tasks automatically.

As we expected, the task completion time – the time the people needed to solve the puzzle – was significantly faster in the video condition than in the audio-only condition (Hypothesis 1). This result could be seen in the automatic analysis using the automatically recorded data. Hypothesis 2 could not be answered as clearly. There were some tendencies showing the video condition to be more enjoyable and that people could express themselves a bit better, but those values were not statistically significant. The data was taken from a questionnaire handed out after the experiment. Detailed results are shown in Gericke et al. (2011).

Fig. 8 Remote study participants in the video condition. Deictic gestures are easily visible for the remote partner

Table 2 Analysis methodologies for the evaluation of design activity (n.y.i. = not yet implemented)

Measure	Manual, questionnaire	Video coding	Automatic analysis
Social measures, emotions	Yes	Partly	No
Whiteboard activity	No	Partly	Yes
Referencing, gestures	No	Yes	n.y.i.
Verbal interaction	No	Yes	n.y.i.
Performance, time-consumption	Yes	Yes	Yes

With this setup, we achieved our goal of showing how using an infrastructure such as the Tele-Board history can ease design team analysis. Capturing and archiving every single event of the communication process can enable many types of analyses conducted instantly without any time-consuming video coding of all experiments. We showed that Tele-Board not only allows asynchronous interaction based on the history of the communication, but also analyses of the design team behavior.

3.3 Connecting Experts on a Sustainability Conference: The B.A.U.M e.V. Jahrestagung

It is Tele-Board's goal to improve collaboration and communication for teams working at different locations without making them travel long distances. Of course, one reason is the cost aspect, but nowadays it is equally as important to avoid traveling due to ecological reasons. Especially when teams have to take long flights to the other location, the "green consciousness" of a company gets into trouble. But not only the CO_2 emissions are problematic; it is also stressful to travel on a long-haul flight including time shifts interfering with the traveler's biorhythm.

As Tele-Board addresses these issues in the broader scope of sustainability, we were invited to present the possibilities of remote collaboration and communication at the annual meeting of the "Federal German working group for eco-friendly

Fig. 9 Results of the "networking table" with Tele-Board between the B.A.U.M. e.V. annual conference in Hamburg and experts in Potsdam, Germany

management" – the *B.A.U.M e.V. Jahrestagung 2011* in Hamburg, Germany. The attendees meet in order to exchange new ideas and develop future concepts together. Applying the idea of "Netzwerktische", i.e. networking tables, groups of about ten people meet at a round table. At each table there is a moderator and an expert for the topic that will be discussed at the table.[4]

At the "Tele-Board table" the expert was located in Potsdam and was supposed to give input from this remote location. In Hamburg, we noted the most important findings from the discussion, sorting and clustering were carried out at the Potsdam location. In addition to the keyboard written notes from the web portal, the participants could scribble and draw their ideas on iPads and send them to the board. Once in a while, the expert summed up the current state of the discussion and the clustering at the whiteboard. When he was doing this, the participants in Hamburg could see his gestures with regard to the content on the board.

At the end of the "network table rounds," all groups had to present the results of their discussion in a 2 min pitch at a podium. At the other tables, the organizers took pictures of the paper notes walls and presented them. The work at our table was presented with whiteboard screenshots downloaded from the Tele-Board web portal as well as pictures showing the work in front of the board (see Fig. 9).

In general, we were enthusiastic about how well the distribution of tasks for this discussion round worked out. It was helpful to have an expert in Potsdam grouping the ideas while we could focus on the discussion. Although the main discussion took place in Hamburg, the person in Potsdam could add his understanding of the discussion by clustering the content. For the participants, it was interesting to see how the sticky notes were moved around and how the different fields of the topic were outlined. In the end, we had a well-structured whiteboard that could be

---

[4] For more information see: http://forum-e3.org/de/enact_2020/idee/ (in German).

presented to the audience. A lesson we learned for a future remote discussion is to improve the audio connection. At this time we had only used one microphone inside the webcam for the whole group in Hamburg. Therefore, it was sometimes necessary to repeat what was said for the Potsdam side. Having a better external microphone or even several for a larger group would ease the understanding between remote partners.

3.4 Network Performance

In contrast to the other presented tests and evaluations, we also did a review of our system from a particularly technical perspective which we presented in Gericke and Meinel (2011). We wanted to show, how the number of users simultaneously connected to the system influences the system's performance. Therefore, we created two setups:

1. Many clients in one synchronized session with variation in the number of clients connected to this session.
2. A fixed number of clients connected to each session with variation in the number of sessions.

To make results reproducible and limit statistical spread within the data, we used a command line client controlling the whiteboard client, which changes a set of simultaneously connected whiteboards periodically and thereby produces a certain amount of activity on the server side. The measurement data revealed that performance characteristics are very different between those setups. Whereas in the first condition load rises exponentially to the number of connected clients, because every client syncs each change to every other client, it is a linear growth in the second setup. Detailed results can be found in Gericke and Meinel (2011).

Looking at the current working mode with the Tele-Board system, teams usually consist of two locations, rarely more. This is because the number of connected whiteboard clients turns out to be more realistic in setup two, so that we can assume our system to scale well in a real-world scenario. The limiting factor on a server is more likely to be the network bandwidth than the computing power. Our numbers also show that the continuous storage of the communication data – enabling asynchronous operation – only has a small influence on system performance. The particularly novelty of our system combining synchronous and asynchronous working modes into one system was shown to be technically feasible.

4 Tele-Board in Use: One Location Setups

In general, Tele-Board is intended to be used in remote location setups. However, in order to improve the usability and workflow of the system, we wanted to gain some insights through single location usage, as well. In the following, we give an overview of two studies showing the use of Tele-Board by students from the School of Design Thinking in Potsdam.

4.1   Physical Versus Virtual Boards: A User Study on Navigation Between Panels

Most of the interactive whiteboard hardware that is available on the market has a resolution of 1,280 × 960 pixels maximum. Compared to traditional whiteboard, this is not a lot of space and people will use several panels, sometimes even simultaneously. Certainly it would be nice to have multiple digital whiteboard devices, but with regard to space and costs it is also possible to switch between two panels on one board. Therefore, we conducted a study evaluating the differences of two displays over one display running two panels.

In this study, the participants were asked to cluster 49 sticky notes into meaningful groups, in order to deduce the most important insights. In one condition, two touch board displays were set up next to each other. One was filled with the sticky notes, and the other was blank. In the other condition, two panels were accessible on a single touch board, one showing the initial sticky notes and the other blank. To switch between them, users tapped a button on the bottom left of the display, which contained a miniature snapshot of the other panel's content.

Results show that working under the restrictions of a single display required slightly more time, yet workflows could continue. Users accepted the visual restriction as a condition of working with a digital system. Team members were also compelled to work more closely together, which both helped and hurt collaboration (for more information see LoBue et al. (2011)).

4.2   As Intuitive as Pen and Paper?: A School of Design Thinking One-Day-Challenge with Tele-Board

Pen and paper is easy to use – for everyone. No matter which professional or cultural background people have, they know how to write an idea on a sticky note. This is one of the reasons why in Design Thinking sticky notes play such an important role. But not only sticky notes are easy to use, working with whiteboards, and also prototyping with material – most people know from kindergarten – is not a great challenge to be learned. With digital tools it is more difficult. Though some people – especially younger ones – love to get to know new devices and tools, for others, it is a burden to have to learn new functionalities and systems.

For this reason, we were eager to know whether and how users were able to work with Tele-Board. Would they manage to do a complete design challenge with our system, only with a small introduction to its functionality? Would they need more

Fig. 10 Setup of the One-Day-Challenge with Tele-Board. At the table and in the hands of the participants are the different tools to write sticky notes

time, compared to working with paper notes and traditional whiteboards? On the other hand, would there be advantages of Tele-Board to help teams to work more efficiently than they are used to?

Therefore, we gave five teams of four people each a design problem and let them work with Tele-Board on a One-Day-Challenge. As we were interested in a comparison with the analog world, we had one additional control team that was working without Tele-Board. All participants had experience with Design Thinking using traditional whiteboards and tools.

In the beginning, we explained all functions of the Tele-Board whiteboard client to the teams and showed them how to write sticky notes with the different devices: We provided four iPads with the *Sticky Pad HD* App (including special iPad pens), a TabletPC, a digital pen (connected to a laptop) and a laptop for writing sticky notes via the Tele-Board web portal (see Fig. 10). The participants were also given a limited amount of time to try out all functions and to get used to the system.

As a main result regarding the time, we found that all teams could accomplish the task and came to satisfying results during the given time frame. We could not see any difference in the timing of the different phases between the control team working without Tele-Board, and the other teams. We could observe that the ease of use and comfort with the system was related to general openness and curiosity towards new technologies and digital tools. That is to say, participants who tried out all Tele-Board functions enthusiastically in the beginning also learned the functions much faster.

Comparing Tele-Board and the traditional tools, we observed that, in general, the teams' usual way of working did not have to be changed and the teamwork was similarly fine. For some people, there was hardly any noticeable difference between traditional tools and the digital system. They even claimed it to be timesaving compared to the analog ones. On the other hand, some participants had difficulties getting used to the system and said it would slow down their work. This was mainly observable with people who were rather shy with trying out all functions. When they could not find what they were looking for in the first place or the system did something they did not expect, they were afraid to try out other things afterwards. Still, with all participants we saw a fast learning effect during the course of the testing. We also observed that it was a great advantage when two of four team members walked through the system easily, because they then showed the others what they found out and after a short while the whole team had no difficulties anymore. In teams where all participants were rather cautious, it took them a longer time to get used to the Tele-Board system.

Besides observing how Tele-Board was used by the teams, we learned which functions were easy to use and which were a bit cumbersome. Thereby, we could also improve the usability of the whiteboard client and add the new functions the teams had suggested. The general feedback of all teams was that they could definitely imagine using Tele-Board for other Design Thinking activities, especially when they have to work in globally distributed teams.

5 Outlook and Future Work

In this chapter we have introduced the Tele-Board system that supports collaborative work in synchronous as well as asynchronous and remote as well as co-located settings (Gumienny et al. 2011). Besides enabling the collective work on the same content at the same time, the automatic storage of whiteboard interaction offers a second-by-second history view (Gericke et al. 2010). The history view enables users to reproduce the progress of a (design) project process. This is very important especially for teams that cannot work on a whiteboard at the same time. Team members can comprehend easier which decisions their colleagues made, earlier in a different time zone.

Recently, we had the opportunity to test Tele-Board in several real world scenarios. In the course of these tests, we were able to monitor user interaction with the system in co-located as well as in remote settings. Our user's general impression of Tele-board was overall positive. Nevertheless, we got valuable feedback that helped us to improve our system. Furthermore, it turned out that for traceability of the design session progress it would be very helpful to recognize the most important phases and present them in the history browser. A prerequisite for suggesting important points of a design session history is an analysis of the stored history data and identifying situations with high information value. Such moments can be, for example, when a team came to certain decisions or had seminal ideas. During the past project year, we collected several hours of test data, which will be the basis for our future research. The obvious commonalities in these processes can be transferred to computational analyses.

5.1   Evaluating Our Ideas and Designs

In the future, we can make use of Tele-Board's advancements of the last years and deploy the system in industry contexts. This way we can conduct long-term research in real working environments on the use of digital whiteboard systems that has not been possible before. We can find out which functions and properties of the system foster and which ones hinder remote collaboration. As Tele-Board can be used with a variety of different hardware, we can easily change settings and try out different setups. Through former tests and experiences we know that we have a good basis for a successful launch of the system and can then adjust it for an optimal experience of the users. Additionally, we will further develop and evaluate special functions for synchronous and asynchronous work. We can demonstrate the possibilities of an all-digital solution such as the Tele-Board system and evaluate the impact for team interaction, their performance and Design Thinking in general.

References

Brown T (2008) Design Thinking. Harvard Business Review (June 2008), pp 84–92

Cross N (2007) Designerly ways of knowing. Birkhäuser, Architektur

Gericke L, Meinel C (2011) Evaluating an instant messaging protocol for digital whiteboard applications. In: Proceedings of international conference on internet computing, CSREA Press, Las Vegas, Nevada, USA, pp 3–9

Gericke L, Gumienny R, Meinel C (2010) Message capturing as a paradigm for asynchronous digital whiteboard interaction. In: Proceedings of the 6th international conference on collaborative computing: networking, applications and worksharing (CollaborateCom), Chicago, Illinois, USA, pp 1–10

Gericke L, Gumienny R, Meinel C (2011) Analyzing distributed whiteboard interactions. In: Proceedings of the 7th international conference on collaborative computing: networking, applications and worksharing (CollaborateCom 2011), IEEE Press, Orlando, Florida, USA, pp. 27–34

Gumienny R, Gericke L, Quasthoff M, Willems C, Meinel C (2011) Tele-Board: enabling efficient collaboration in digital design spaces. In: Proceedings of the 15th international conference on computer supported cooperative work in design, CSCWD '11, IEEE Press, Lausanne, Switzerland, pp 47–54

LoBue P, Gumienny R, Meinel C (2011) Simulating additional area on Tele-Board's large shared display. In: HCI International 2011 – Posters' Extended Abstracts, Springer, Berlin/ Heidelberg, pp 519–523

Lockwood T (2009) Design thinking: integrating innovation, customer experience, and brand value. Allworth Press, New York

Martin RL (2009) The design of business: why design thinking is the next competitive advantage. Harvard Business Press, Boston

Plattner H, Meinel C, Weinberg U (2009) Design thinking. mi-Wirtschaftsbuch, Munich

Applied Teamology: The Impact of Cognitive Style Diversity on Problem Reframing and Product Redesign Within Design Teams

Greg L. Kress and Mark Schar

Abstract In the words of Professor Larry J. Leifer, "All design is redesign." As designers collect information about a problem, they form a mental frame of the problem space that is the scaffolding around which to build a solution. When presented with new information, successful designers can "reframe" the problem and the solution as part of a successful iterative cycle. These iterative cycles are central to the Stanford Design Thinking process. A team's capacity and willingness to reframe can be measured by means of a closed-form assessment tool that eliminates many of the confounding variables of the previous longitudinal (project performance-based) approach. We propose the Stanford Design Thinking Exercise (SDTE) as a measure of reframing behavior and design team effectiveness. The exercise is standardized and can be conducted in a controlled lab or classroom setting in 1 h. The SDTE is designed to be a first step toward a quick, reliable and standardized technique for evaluating design team effectiveness. We found that the SDTE is a robust measurement for reframing change, in that it reports a range of reframing results across a participant population group, but attempts to align the instrument with participant cognitive characteristics were unsuccessful indicating that more work needs to be done to understand specific indicators of reframing.

1 Introduction

The activity of design follows an inherently nonlinear process. In design, it is rarely the case that a specific design solution can be reached through a linear combination of independent components, conducted in lock-step process order. The design process is often represented as a series of steps that appear linear (see Fig. 1). In fact, it is widely assumed that the design process is iterative in nature and often involves a "back-and-forth" between the principal steps within a prescribed process.

G.L. Kress (✉) • M. Schar
Center for Design Research, Stanford University, 424 Panama Mall, Stanford, CA 94305, USA
e-mail: glkress@stanford.edu; mfschar@stanford.edu

H. Plattner et al. (eds.), *Design Thinking Research*, Understanding Innovation,
DOI 10.1007/978-3-642-31991-4_8, © Springer-Verlag Berlin Heidelberg 2012

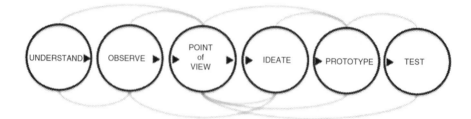

Fig. 1 Steps in a Design Process (Ratcliffe 2009). This six-step process is widely regarded as the standard as applied and developed at the Hasso Plattner Institute of Design at Stanford. Of particular interest is the interconnection between the steps and the backward looping process that suggests both iteration and reframing

Bruno Latour, a French sociologist and anthropologist who is influential in the study of science and technology has stated, "to design is always to redesign." (Latour 2008) Latour suggests that the design process never begins from scratch, that designers begin with a specific cognitive framing of a problem and then redesign toward "something more lively, more commercial, more usable, more user friendly, more acceptable, more sustainable and so on, depending on the various constraints to which the project has to answer."

This would indicate that it is important to understand the "design thinking" involved in the act of redesign as a central component of the design process. Presented with a problem or issue within a specific mental framework, how does the designer deconstruct this framework and construct a new framework toward a novel solution? When presented with new information, how does the designer revise his or her previous approach? This also suggests that certain cognitive capabilities may be more valuable or useful in the process of redesign. If these cognitive capabilities can be identified and defined it would be a significant step forward in the understanding of design thinking.

2 Experiment Overview

This experiment aims to model team cognition and diversity in order to understand how certain cognitive characteristics in a team can influence team dynamics and performance. Nearing the end of its second year, this study has produced encouraging and significant results that merit further study, as well as building a platform for international cooperative research in the context of the ME310 Global Design Innovation course community.

2.1 Pilot Study and Limitations

The initial 2009–2010 "Teamology" study observed existing student design teams within the context of the ME310 Global Design Innovation course (see Fig. 2).

Fig. 2 Pilot study (2009–2010)

In this course, students work for 8 months in teams (which are often distributed between two different universities) to solve engineering problems presented by a corporate client. Multiple cognitive assessments were performed on all the students who participated ($n = 97$), and these assessments were used to create a cognitive model of each team. We then evaluated the quality of their project outcome at the end of the course using a 7-point scoring metric pre-adjusted for project difficulty. Using this method, we were able to confirm that there is substantial cognitive diversity in the ME310 student population, even where ethnographic diversity is not evident. We were also able to find at least one strong and significant correlation between a team's cognitive profile and their performance outcome, suggesting that further exploration is merited.

However, there were several limitations to this research method. First, we had no control over forming teams, and so we could only observe the cognitive variation that was inherent in the population. Thankfully, this was substantial enough to perform our analysis. However, it did not give us the degree of control necessary to perform rigorous tests of our hypotheses. Additionally, various types of diversity are present at one time, making it very difficult to parse the effects of different cognitive characteristics. Since all the projects are unique, it is also difficult to perform a reliable performance assessment and comparison on which to draw conclusions. Lastly, the 8 month duration of the project and the internationally distributed subject pool make this a challenging and lengthy analysis.

2.2 SDTE Approach

Ideally, we would like to assess design performance in the short term and in a controlled environment. This would allow us to directly and reliably test the impact of particular cognitive characteristics, but it is not limited to this purpose. A reliable and quick team assessment tool could have major implications for team-based education and design in industry. The Stanford Design Thinking Exercise (SDTE) is an attempt to create a tool that can measure one indicator of design team performance (reframing behavior) in less than 1 h. The exercise is standardized and mostly closed-form, so it can be reliably and quantitatively scored. Ad hoc and pre-existing teams will work through the exercise in a controlled laboratory setting and will follow identical sets of instructions, eliminating several confounding variables (see Fig. 3).

In developing this exercise, we can create teams with known cognitive profiles in order to directly test our hypotheses. We can measure if reframing behavior is

Fig. 3 SDTE approach (2010–2011)

Fig. 4 Full experiment (2011–2012)

affected by certain cognitive characteristics, and we will have a basis for team comparison that is far less subjective and time-consuming. Researchers have had success in video coding of short-term team interactions as an indicator of long-term performance, though these methods are also very time-consuming and resource-intensive. Scoring of the SDTE is simple enough to be programmed into an Excel spreadsheet.

2.3 Full Experiment

In the complete experiment, a combination of test and control teams will be formed within the global ME310 course (see Fig. 4). Local and international participation will be extended to the greatest extent possible. Teams formed along particular cognitive guidelines will comprise the test group, and the control group will be formed at random. Teams will then be assessed on the long-term performance outcome of their design projects, as well as in the short-term by means of the SDTE. This will allow cross-validation of the SDTE as a design performance measurement tool as well as of our team cognition hypotheses.

3   Framing, Reframing and Design Team Composition

3.1 Framing

A frame, in social theory, is a schema of interpretation, the method by which an individual organizes experience and knowledge to solve problems. Erving Goffman described frame analysis as the "understanding available in our society for making sense out of events and to analyze the special vulnerabilities to which these frames of reference are subject." (p. 10) (Goffman 1974) Goffman describes framing as an examination of how an individual organizes experience, how they subjectively structure involvement in an event or circumstance. Simply stated, Goffman believes the frame analysis is the "way we take it that our world hangs together." (p. 440)

Designers bring their individual frames to the problems of design. These range from simply an understanding of "the way the world works" to more sophisticated perspectives around functionality and performance based on prior experience or expertise. If, indeed, "all design is redesign," then the ability to reframe, or break apart existing frames and reassemble new frames of understanding becomes a critically important activity within the context of design thinking.

3.2 Reframing

We define "reframing" as the disassembling of an existing problem solving frame set and reassembling a new frame set, possibly leading to a new problem solution. This definition has its roots in neuro-linguistic programming, a psychotherapeutic technique designed to change patient behaviors through reinterpretation of subjective experiences. (Bandler et al. 1982) This is also similar to work in cognitive linguistics, particularly the work of George Lakoff, which examines the shifting public opinion on political issues through language that promotes reframing of issues. (Lakoff 2008)

We are also interested in the designer's propensity to reframe; their willingness to shift an existing problem-solving frame set based on the introduction of new information. In some cases, the new information can be important to shifting the problem-solving frame set, in which case reframing is an appropriate activity. In other cases, the new information may not be important and reframing the problem solving set may not lead to a better answer. The designer is responsible for deciding how this new information will (or will not) contribute to the reframing of the problem and solution.

3.3 Implications for Design Team Composition

This propensity to reframe may be tied to a specific cognitive problem solving style. For example, it is been shown that teams with higher scores on the personality factor of "openness to experience" as measured by the NEO-FFI psychometric instrument (McCrae and John 1992) perform better on team creativity tasks. (Schilpzand et al. 2011) However, in this same study, teams that were diverse on "openness to experience" had the highest level of measured team creativity, as long as they have some team members that are low on openness and others that have a moderate level of openness to experience. This would suggest that team diversity in reframing abilities might be more critical to team creativity than teams composed of individuals with higher reframing abilities.

Additionally, our previous research on the cognitive characteristics of ME310 design project teams has offered strong evidence that the presence of the cognitive mode "extraverted feeling" (EF) as defined by the Wilde-Type "Teamology" method correlates strongly with the quality of the team's design outcome. (Wilde 2008; Kress and Schar 2011) It is worth investigating if the presence of EF also appears to influence reframing behavior both at the individual and the team level. We also hope to explore the impact of cognitive problem-solving styles and propensity to reframe on team dynamics. Team reframing may be moderated by specific individuals who lead the reframing process within a team, similar to "group contagion" identified in research on the impact of a "bad apple" on group dynamics. (Online survey software tool – create free online surveys – Zoomerang) If this were the case, it would have a significant impact on design team composition.

4 Development of the SDTE

4.1 Background

The objective of the Stanford Design Thinking Exercise (STDE) is to measure reframing on both an individual and group level. To translate a mental frame into observable behavior, subjects express their initial frame set through a series of individual choices. In this manner, a frame is described as a unique collection of choices that solve a design problem. This concept is known as "choice structuring" and has long been associated with organizational learning and strategic business planning. (Artiz and Walker 2010)

Argyris distinguishes choice structuring as either "single loop" or "double loop" learning. In single loop learning, choices are operationalized rather than questioned in an effort to most efficiently move from beginning to conclusion of a problem. In double loop learning, choices are linked in a cascading structure in each choice may be questioned in terms of governing variables or consequences for proceeding to the next choice. In a double loop learning choice structuring process, error detection (or disagreement in a group context) is corrected immediately through modification of an individual's or group's underlying norms or objectives. (Online survey software tool – create free online surveys – Zoomerang) The STDE was designed to be a double loop learning choice structuring instrument that could simultaneously measure both an individual and group reframing of a potential problem solution.

4.2 Case Study Approach

This instrument is an episodic case study that requires both individual and group choice structuring. The subject of a case study was an "urban public share bicycle," where subjects are asked to design a bicycle suitable for an urban environment and shared by the general public. This topic was chosen because it struck a balance of familiarity; most subjects are familiar with bicycles, but generally not in a shared public context.

The case study involves a short paragraph describing the situation and presenting the subject with 10 initial design choices. The subject is required to rank those design choices from 1 (most important) to 10 (least important). This process of choice structuring within a case study has been used in other research instruments, particularly involving survival scenarios, with titles such as "Lost at Sea" or "Lunar Survival." For example, a similar case study based choice structuring process was used to identify decision making within intercultural teams. (Artiz and Walker 2010)

The case study instrument is designed to be episodic with a series of "rounds" where new information is introduced and new choice structuring (or reframing) can be measured. The STDE included three rounds which in total present the subjects with 20 items for choice structuring. A fourth round offers the opportunity for additional choice structuring in a more open format. A complete discussion of the STDE instrument and application is found in Sect. 5: SDTE in Application. A complete example of the instrument as distributed to subjects is available in Appendix 8.1.

4.3   Item Definition and Validation

The first step in developing items for subject ranking was to brainstorm a list of potential design choices which fell into two basic groups – object-oriented or user-oriented. The reason for this distinction was to identify items that were significantly distinct and offered a greater range of choice when included as a design option. Additionally, the object- and user-oriented choice groups represented two implicit "strategies" for the design solution.

Object-oriented items were design choices that are associated with specific, physical aspects of the bicycle design and included items like "The tires are textured and flat resistant" or "The frame is made from thin-walled high grade aluminum." User-oriented items were design choices that are associated with the user experience with the bicycle and included items like "the bicycle is easy to ride" or "the rider sits in a comfortable upright position." A total of 31 items were developed (15 object-oriented and 16 user oriented) and tested with an independent subject group for a match with either a provided definition of object or user orientation. Items with less than 85 % agreement with the appropriate definition were discarded, resulting in a final 10 object-oriented and 10 user-oriented items for testing. In further iterations, the items were refined to include an assortment of decidedly object-oriented, user-oriented and ambiguous items. The final list of 20 items can be found in Appendix 8.2.

The next step was to determine the order of importance of each of the items. A second independent sample of subjects were presented with the urban bicycle design case study and asked to rate each of the items in order of importance to the design challenge on a five-point Likert-like scale from 1 (very important) to 5 (not important). This resulted in a general consensus on the order of importance of each item within the task of designing a successful urban-shared bicycle. Statistical t-tests showed a significant difference in rating for items separated by about three positions on the rating scale meaning that, for example, the item ranked #7 is significantly different than the item ranked #10, as shown in Appendix 8.2.

This item ranking allows for the presentation of items to subjects in a known order of importance (ascending or descending) over subsequent rounds. Items presented in an ascending order of importance will likely result in maximal change to the choice structure decisions (maximum reframing) while items presented in a descending order would likely result in minimal change (minimum reframing).

4.4 Item Validation: Mechanical Turk

For item validation we used the internet-based research tool Mechanical Turk by Amazon.com to solicit subjects. Mechanical Turk (MT) offers a quick, cost-efficient way to source potential subjects but it also has disadvantages that need to be addressed in data processing. MT offers compensation to potential subjects in the range of $.25–50 per survey experience which translates to $4–5/h, making the service attractive to both legitimate subjects and internet scammers.

We use MT to solicit subjects and then link those subjects to a professional quality survey tool called Zoomerang by MarketTools. (Online survey software tool – create free online surveys – Zoomerang) The Zoomerang online survey tool collects all the data and at the conclusion of the survey offers the subject a completion code. The subject automatically returns to MT, enters the completion code and then receives credit for completing the task. We restricted participation in MT to US residents who have a 95 % or better performance rating and have participated in no more than two of our previous surveys. This offers some protection, though scammers can create fake accounts or possess multiple accounts to bypass this screening mechanism.

We use two methods (built-in to MT) to identify those subjects who have dutifully completed the survey instrument: duration of task and click stream analysis. For duration analysis, we know the total time a subject used to launch the MT task, complete the Zoomerang survey and return to MT. We calculate a total subject mean time for completion and establish a two-degree standard error of the mean (SEM) range. Subjects that fall outside of this range are excluded from the data set; this is typically 10–15 % of subjects.

The second method we use to screen subjects is a click stream analysis designed to identify subjects who picked the same response for each question in an effort to complete the survey quickly. Each survey is constructed with at least one-third reverse-scored questions. For each question, we establish a mean score and a one-degree standard error of the mean range. If the subject has more than 50 % of their item responses outside of the one SEM range, which typically translates to less than a 1 % overall probability, then we discard this subject. We also review the responses from each subject that is discarded by this method to confirm that the content of the response does in fact seem suspect.

Through both the time of completion screen and the click stream analysis we typically discard 25–35 % of subjects who complete the MT task. While this likely means that we have screened out some legitimate subjects, it also ensures that we are screening out 95 % of the behaviors that are statistically extreme and thus suspect.

5 SDTE in Application

5.1 Explanation of the Exercise

The SDTE is a team exercise comprised of four rounds, each with an individual and a group component. The team is asked to role-play as a design team within a company that makes bicycles. They are tasked with designing an "urban public share bicycle" by selecting and ranking features, components and design elements from a list provided to them. In each subsequent round, additional options are provided to the group and they are asked to revise their previous ranking. They do so first as individuals, and then come together to reach a group consensus. In the final round, participants are given an open option to create up to five new options (in contrast to the previous three rounds, where all information is provided and fixed). In total, the four individual and four group rounds of the exercise provide $4(n + 1)$ cases that can be scored independently, where n represents the size of the participant group.

The four rounds of the experiment each contain new feature options, creating the opportunity for the team to revise their previous solution (reframing). The first round, or Initial Set, provides ten items to be ranked. The second round, or Stimulus Set, offers a new set of five options from which the top ten are to be chosen and ranked. In the third round, the group is given the Final Set which completes the pre-defined list of 20 items. They are then asked again to select and rank the top 10. In the fourth round, the participants are asked to provide their own new feature options. They may create up to $5n$ new options during the individual round from which a group set of up to five are chosen. Individuals and groups may choose to create fewer than five or none at all. As such, the final set is anywhere from 20 to 25 of which the top ten must be chosen and ranked (see Table 1).

5.2 Scoring the Exercise

At the end of each round, each individual writes a rank ordering of ten items into the score sheet (see Appendix 8). Scoring is accomplished by tallying the amount of change from one round to another. Change is measured by the number of spaces in the score chart that a given item moved from one round to another. For example, if

Table 1 Items sets by round

Round	Set name	New options	Cumulative options	Choice set
1	Initial set	10	10	10
2	Stimulus set	5	15	10
3	Final set	5	20	10
4	Open set	5[a]	25[a]	10

[a]The group may create up to five new options, but are not required to create any.

an item moved from slot 2 to slot 5, this would be scored as a change of 3 (no sign). The changes for items that are newly added to the list are scored as if they were brought up from the bottom of the ten-item list (think of it as slot 11); for example, if an item is newly in the ranking at slot 3, this would be scored as a change of 8. Items that are not moved or that are removed from the ranking are not scored. Finally, the ten change scores are summed up to give a total reframing score. This process is fully automated by means of a MATLAB scoring algorithm such that analyzing a given trial takes a few minutes at most. Furthermore, the simplicity of this scoring rubric allows for complete trials to be permanently recorded in Excel format with no loss of information.

Because of the structure of the scoring scheme, minimum reframing scores for all rounds are zero. Maximum scores for reframing are 50 for group rounds (where no new information is introduced) and 65 for individual rounds (where 5 new items are introduced per round). This range allows us to understand where individuals and teams fall on a continuum of reframing behavior. Additionally, the exercise allows for multiple different types of scoring:

5.2.1 Individual-to-Individual Comparison

Individuals on the same team exhibit different amounts of reframing from round to round. This would allow us to investigate if individual cognitive characteristics contribute to personal reframing behavior. To do this, individuals are compared on the amount of reframing they do during individual rounds.

5.2.2 Individual-to-Group Comparison

Individuals must change their solutions to different extents to match the group consensus. Often, a strong personal presence can dominate the discussion and force a particular group consensus. This is apparent when individuals have very imbalanced amounts of reframing relative to the group consensus. To do this, individuals are compared on the amount of reframing they must do to match the group consensus in each round.

5.2.3 Group-to-Group Comparison

Different groups will reframe differing amounts when presented with the same exercise. This difference is evident in the scoring of the exercise. Groups can be compared either by the average of all individual reframing, as the average overall reframing between group consensus rounds alone, or both.

5.2.4 Group-to-Consensus Comparison

Since the items have previously been ranked in importance by a relatively large group of independent participants, we have some sense of the general consensus as to the importance of these items to the final design (see Appendix 8). As such, there is the expected outcome that at the end of each round the ranking will simply be the top ten ranked items of the given item set in descending order. Particular group consensus scores can then be compared on their overall difference from the expected outcome.

5.2.5 Additional Opportunities

The open response format of the fourth round allows for additional coding opportunities and other scoring approaches (such as counting the total number of ideas generated, whose ideas comprised the consensus, were the ideas reformulated in paring down, etc.).

5.3 Tuning the Exercise

In order to get meaningful variation from the experimental group, it is necessary that the exercise allow and encourage enough reframing in such a way that teams predisposed to reframe will do so. However, the stimulus must not be so strong that it forces reframing behavior on teams that would otherwise not act in such a way. As such, we would like to tune the exercise to set some medium level of expected reframing. We can do this by using the general consensus information to regulate when salient pieces of information are delivered to the team.

For example, a team for whom highly relevant items come in the last round is likely to exhibit more reframing behavior in the last round. The opposite is true if all highly relevant items are given in the beginning. Of course, there is a very large number of permutations in how the 20 items can be subdivided into 10, 5 and 5. The general consensus rankings give us a way to measure how much reframing is expected for each of those permutations. Tuning of the exercise has been fully automated by means of a MATLAB algorithm that generates an appropriate permutation for whatever size stimulus is desired.

6   Results and Analysis

Altogether, we performed approximately 30 trials of the exercise for pilot testing and validation. 12 of these trials were during the final refinement and calibration of the current version. The following 18 trials were given with the final version of the exercise tuned to two different levels of stimulus and were rigorously scored and analyzed together with subject cognitive style data.

6.1   Pilot Test Results

The first dozen trials of the current version of the exercise were tested with random groups of undergraduate and graduate students. During these trials, we tuned the exercise to minimum, maximum and several intermediate levels of expected reframing. We initially found that individuals and groups were both responsive to the size of the reframing stimulus (i.e., the relative importance of incoming information) in a fairly consistent pattern (see Figs. 5 and 6), and also that there is substantial variability between different individuals and groups. This is what we had hoped to see, as we would like to have sufficient tuning control over the size of the stimulus and the expected outcome, and also that at a given tuning there is still sufficient variability to make cross-team comparisons. These results appear consistent regardless of whether one is looking at the individual or group level, at only one round or across many (for additional results, refer to Appendices 8.3 and 8.4). Additionally, each group's end result was distinct, allowing a final comparison of the team's design "solution" against other teams', and against the general consensus. These early trials seemed to indicate that the scoring and tuning metrics were robust.

6.2   Validation Trial Results

The 18 trials that followed all used the final version of the exercise, calibrated to a total reframing stimulus of either 63 or 84. They were given to 18 different teams comprised of three individuals each (in one case there was a team of four), totaling 55 participants. We find that there is not a significant difference in total individual or group reframing at these two different levels of stimulus. In the stimulus 63 case, 30 participants exhibited on average 63.2 total reframing actions, with a standard deviation of 11.0. In the stimulus 84 case, 25 participants exhibited on average 64.4 total reframing actions, with a standard deviation of 13.3. In the stimulus 63 case, group reframing was on average 56.4 total actions, with a standard deviation of 12.8. Groups in the stimulus 84 case exhibited 62.2 total reframing actions on average, with a standard deviation 9.9. We can see that this result is much closer to a significant difference, though it is still impossible to reject the null hypothesis. We also find that we cannot confirm a normal distribution in the individual or group reframing scores.

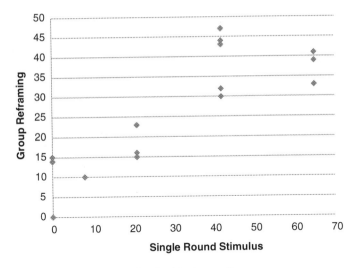

Fig. 5 Group reframing response to stimulus (single round)

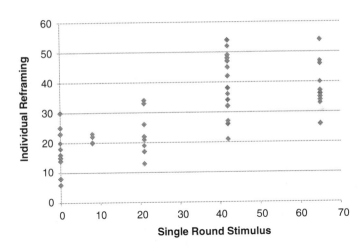

Fig. 6 Individual reframing response to stimulus (single round)

6.3 Cognition Results

We find that there is no significant correlation between individual reframing behavior (adjusted for stimulus) and any of the eight Wilde cognitive modes, including a general measure of Extraversion. We examined this from several scoring perspectives; first, we considered total individual reframing on the exercise calculated as a sum of the reframing activity between subsequent iterations of that individual's solution. Second, we examined the amount of reframing that each individual did immediately after a group consensus (and after receiving the stimulus). This could be used to measure the direct individual response to stimulus and also the propensity of individuals to rework the group solution. Third, we examined to what extent the group consensus in each round was different from the previously submitted individual rankings. This measure could be an indicator of domineering behavior, as dominant individuals tend to mold the group decision to reflect their own. None of these scoring metrics showed any correlation with Wilde mode scores; the strongest correlation was between the general Extraversion measure and post-consensus individual reframing ($r = 0.234$, $p = 0.088$, $n = 54$). This is likely a statistical artifact. There were no direct correlations with mode scores at a level of significance above 90 %. Group-level cognitive measures such as team mode means and cognitive diversity did not show any significant correlations with group-level reframing behavior; other team characteristics such as gender composition also did not affect reframing significantly.

6.4   Accuracy Results

In this exercise, accuracy is defined by the extent to which an individual's or group's solution adheres to the general consensus ranking of items as determined during the development of the exercise. We find a slight correlation between the individual reframing score (adjusted for stimulus) and individual accuracy on the exercise ($r = 0.31, p = 0.02, n = 55$). However, we find that if a general consensus ranking was to be constructed from these trials, it would not necessarily match the general consensus determined earlier on which this accuracy measure is based. There were no correlations found between group consensus accuracy and any of the group-level measures, including gender composition, cognitive style or cognitive diversity.

7   Conclusions and Future Work

As expected, we found that different individuals exhibit different amounts of reframing behavior during the design exercise. Unfortunately, we were unable to establish any reliable links between this variation in behavior and individual/team-level cognitive characteristics. Though we observed some initial successes in being able to "tune" the amount of reframing by adjusting stimulus size, we find that overall the natural variation in reframing behavior is far noisier than the stimulus response. As such, there is no support for our hypothesis that reframing behavior is a reliable expression of cognitive style, or that more reframing behavior in the field can explain the greater success of certain teams. There is no compelling evidence that the SDTE measurement is as reliable an indicator as we had hoped.

After these somewhat discouraging results, we are forced to reevaluate our approach for the coming year of research. The question remains open as to precisely how cognitive characteristics are manifested in team interactions, and precisely how these interactions can be understood to influence team performance. We will defer future refinement and testing of the SDTE in favor of a three-pronged deeper investigation into the links between cognition, dynamics and design team performance:

- **Expanded data set.** In addition to analyzing our data from the 2009–2010 academic year, there is also the possibility to expand our analysis to include the past five academic years (2007–2008 through 2011–2012) of ME310 students at Stanford. This will be done with existing data sets and archival project records. Expanding our data set and increasing our sample size should offer more clarity and strength to our statistical findings.
- **Behavioral analysis.** Existing data sets from ME310 2009–2010 include extensive records of behavioral observations (e.g. SPAFF) collected by past doctoral students in the Center for Design Research. We are already working on merging these.
- **Direct team observation.** We have the opportunity to explore novel dependent measures by direct observation of the current ME310 student teams. Findings from the prior behavioral analysis and expanded data set will inform the nature of these observations.

8　Appendix

8.1　SDTE

Urban Public Share Bicycle – Round 1

You are the head of a product design team charged with developing a bicycle for a major global bicycle manufacturer who intends to bid on the next generation of "urban public shared bicycles" for the cities of Berlin Germany, Barcelona Spain, and San Francisco USA. If this bicycle design is successful, you expect to sell well over 100 K units in the first year.

Your team has developed a list of potential design options to pursue in future work. Your job is to rank order these design options from what is **most important** to consider to what is **least important** to consider. Please rank order the list below from **1** (most important to consider) to **10** (least important to consider) and place your answers on the **Answer Sheet** under the column labeled **Round 1 - Individual**:

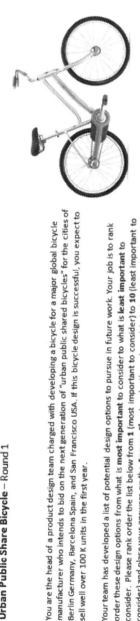

Urban Public Share Bicycle
Example Only: This is not the acutal bicycle

Code	Items
A	The rider stays clean because most mechanisms are covered
B	There are 30 different color combinations to choose from
C	The front and rear axles have low friction bearings
D	The frame is made from thin walled high grade aluminum
E	The tires are textured and flat resistant

Code	Items
F	The tires give the rider a feeling of confidence and safety
G	The company makes Tour de France bicycles
H	The locking mechanism is made of lightweight titanium
I	The bicycle helps the rider feel like that are "living green"
J	The company motto is "classic bikes for a modern era"

When everyone has ranked the items individually, then begin discussing and arrive at a consensus **Group Ranking**. Place your answers on the **Answer Sheet** under the column labeled **Round 1- Group**.

Urban Public Share Bicycle – Round 2

Your team has talked with the various parts suppliers and has discovered five new design options that can be incorporated into this design. These new design options are listed below. Your job is to rank order all the design options (the 10 original and these 5 new design options) from what is **most important** to consider to what is **least important** to consider. Please rank order the list below from **1** (most important to consider) to **10** (least important to consider). You may only rank 10 items and must exclude 5 items. Place your answers on the **Answer Sheet** under the column labeled **Round 2 - Individual**.

Code	Items
K	The rider can easily carry groceries or a briefcase
L	The drive chain is a one-piece break resistant belt
M	The bicycle is easy to ride
N	The rider sits in a comfortable upright position
O	A pitlock locking skewer is affixed to the wheels

When everyone has ranked the items individually, then begin discussing and arrive at a consensus **Group Ranking**. Place your answers on the **Answer Sheet** under the column labeled **Round 2-Group.**

Urban Public Share Bicycle – Round 3

Your team has talked with the various parts suppliers and has discovered five new design options that can be incorporated into this design. These new design options are listed below. Your job is to rank order all the design options (the 10 original, plus 5 items from the last round and these 5 new design options) from what is **most important** to consider to what is **least important** to consider. Please rank order the list below from **1** (most important to consider) to **10** (least important to consider). You may only rank 10 items and must exclude 10 items. Place your answers on the **Answer Sheet** under the column labeled **Round 3 - Individual:**

Code	Items
P	The rear fender has room for a logo and user instructions
Q	The seat adjusts easily with a compression ring and flip lever
R	The back wheel has a carbon steel disc brake
S	The down tube gently curves giving the bicycle a sophisticated and durable look
T	The handlebars are comfortable for the rider

When everyone has ranked the items individually, then begin discussing and arrive at a consensus **Group Ranking.** Place your answers on the **Answer Sheet** under the column labeled **Round 3- Group.**

Urban Public Share Bicycle — Round 4

You now have the chance to add your own design options, if you'd like. If you have any ideas not covered in the previous design options, write them in the space below. You do not have to add any options if there's nothing you would add. Your job is to rank order all the design options (the 15 original from Round 1, the 5 design options from Round 2, the 5 design options from Round 3 and any new options from this round) from what is **most important** to consider to what is **least important** to consider. Please rank order the list below from **1** (most important to consider) to **10** (least important to consider). You may only rank 10 items and must exclude all other items. Place your answers on the **Answer Sheet** under the column labeled **Round 4 - Individual**.

Individual	
Code	items
U	
V	
W	
X	
Y	

Restated By Group for Ranking	
Code	items
AA	
BB	
CC	
DD	
EE	

When everyone has ranked the items individually, then begin discussing the new items and restating them as a group. Then for all the items arrive at a consensus **Group Ranking**. Place your answers on the **Answer Sheet** under the column labeled **Round 4 - Group.**

Urban Public Share Bicycle – Answer Sheet

Round 1			Round 2			Round 3			Round 4	
Rank	Individual	Group	Rank	Individual	Group	Rank	Individual	Group	Rank	Individual	Group
1			1			1			1		
2			2			2			2		
3			3			3			3		
4			4			4			4		
5			5			5			5		
6			6			6			6		
7			7			7			7		
8			8			8			8		
9			9			9			9		
10			10			10			10		

8.2   *Consensus rankings*[a]

Rank	Item	Rank	95%CI
1	The bicycle is easy to ride	1.68	0.26
2	The handlebars are comfortable for the rider	1.80	0.26
3	The tires give the rider a feeling of confidence and safety	1.88	0.24
4	The rider sits in a comfortable upright position	1.96	0.26
5	The tires are textured and flat resistant	1.97	0.24
6	The rider stays clean because most mechanisms are covered	2.15	0.25
7	The rider can easily carry groceries or a briefcase	2.25	0.24
8	The seat adjusts easily with a compression ring and flip lever	2.28	0.24
9	The front and rear axles have low friction bearings	2.31	0.21
10	The drive chain is a one-piece break resistant belt	2.32	0.24
11	The frame is made from thin walled high grade aluminum	2.55	0.20
12	The locking mechanism is made of lightweight titanium	2.66	0.21
13	The back wheel has a carbon-steel disc brake	2.69	0.21
14	The down tube gently curves giving the bicycle a sophisticated and durable look	2.78	0.20
15	The bicycle helps the rider feel like they are "living green"	2.80	0.24
16	A pitlock skewer locknut is affixed to the wheels	2.95	0.20
17	There are 30 different color combinations to choose from	3.18	0.23
18	The company motto is "classic bikes for a modern era"	3.32	0.26
19	The rear fender has room for a logo and user instructions	3.34	0.23
20	The company makes Tour de France bicycles	3.49	0.23

[a]via Amazon Mechanical Turk, $(n = 95)$.

8.3   Group Reframing Response to Stimulus (Total)

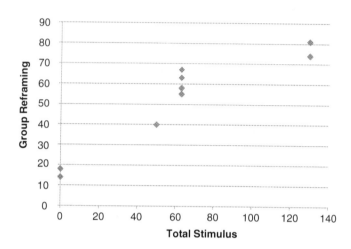

8.4   Individual Reframing Response to Stimulus (Total)

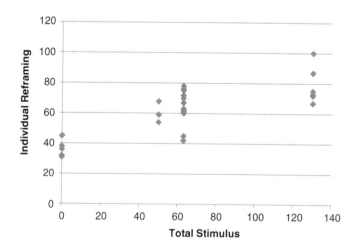

References

Argyris C (1976) Single-loop and double-loop models in research on decision making. Adm Sci Q 21(3):363–375

Artiz J, Walker RC (2010) Cognitive organization and identity maintenance in multicultural teams: a discourse analysis of decision-making meetings. J Bus Commun 47(1):20–41

Bandler R, Grinder J, Andreas S (1982) Neuro-linguistic programming and the transformation of meaning. Real People, Moab

Felps W, Mitchell TR, Byington E (2006) How, when, and why bad apples spoil the barrel: negative group members and dysfunctional groups. Res Organ Behav: Annu Ser Anal Essays Crit Rev 27:175–222

Goffman E (1974) Frame analysis: an essay on the organization of experience. Harvard University Press, Cambridge

Kress G, Schar MF (2011) Initial conditions: the structure and composition of effective teams. In: Proceedings of the international conference on engineering design (ICED), Copenhagen

Lakoff GP (2008) The political mind: why you can't understand 21st-century American politics with an 18th-century brain. Viking Adult, New York

Latour B (2008) A cautious prometheus? A few steps toward a philosophy of design (with Special Attention to Peter Sloterdijk). In: Networks of design: proceedings of the 2008 annual international conference of the design history society (UK) University College, Falmouth, p 13

Martin R (2002) The responsibility virus: how control freaks, shrinking violets – and the rest of us – can harness the power of true partnership. Basic Books, New York

McCrae RR, John OP (1992) An introduction to the five-factor model and its applications. J Pers 60(2):175–215

Online survey software tool – create free online surveys – Zoomerang. http://www.zoomerang. com. Accessed 02 July 2011

Ratcliffe J (2009) Steps in a design thinking process. Stanford HPI d.school K-12 Lab

Schilpzand MC, Herold DM, Shalley CE (2011) Members' openness to experience and teams' creative performance. Small Group Res 42(1):55–76

Wilde DJ (2008) Teamology: the construction and organization of effective teams. Springer, New York/London

Qualitative Methods and Metrics for Assessing Wayfinding and Navigation in Engineering Design

Jonathan Antonio Edelman and Larry Leifer

Abstract Designing can be viewed as a body of behaviors. Fundamental to several design behaviors is Path Determination. Path Determination describes the moments when designers choose what they will take up for development as well as how they experience their perceptual horizon.

Our research suggests that there are two primary modes of Path Determination, Wayfinding and Navigation. Each of these has been correlated with different outcomes in redesign scenarios. Wayfinding correlates to making significant changes to an object, while navigation correlates to making incremental changes to an object.

In this chapter, I present a novel methodology for capturing and observing Wayfinding and Navigation behaviors, as well as several metrics for measuring these behaviors.

1 Introduction

The scope of product design ranges from making incremental improvements to existing products to making radical departures from existing products. At the heart of design, whether incremental improvements or radical redesign, are designers and their behaviors. Path Determination, when and how they choose to follow a course of action, has proven to have an impact on outcomes. A new methodology for assessing the elements of Path Determination and applying meaningful metrics is a necessary first step in crafting a curriculum for high performance engineering design teams.

J.A. Edelman • L. Leifer (✉)
Center for Design Research (CDR), Stanford University, 424 Panama Mall, Stanford, CA 94305-2232, USA
e-mail: leifer@cdr.stanford.edu

H. Plattner et al. (eds.), *Design Thinking Research*, Understanding Innovation, DOI 10.1007/978-3-642-31991-4_9, © Springer-Verlag Berlin Heidelberg 2012

1.1   Redesign

Redesign is a major, and by no means trivial challenge for designers. Redesign tasks provide a good platform for understanding radical breaks as they permit an evaluation of the differences between the model provided at the start, and the newly designed model at completion. In addition, redesign is a prevalent design activity in professional design practice.

1.2   Incremental Redesign

The task of redesign is often a matter of making incremental improvement to an existing design. The process of optimization and incremental improvement is practiced with rigor, and is a remarkable achievement for both industry and the academy. It is in part because of these practices that products are more robust. Ph.D. programs with emphasis on optimization, axiomatic design, and Design for X, are a core part of Engineering Design curricula and are common in Universities around the world.

1.3   Radical Redesign: What Is a Radical Break?

Radical breaks occur in redesign scenarios when a new design is extremely divergent from the original design. All design is redesign, thus an understanding of radical breaks as an extreme case is essential to understand the limits of how design is accomplished.

The notion of radical breaks captures what is often thought of as "thinking outside of the box", and reframing problems to find new and unique solutions.

The process that leads to radical breaks is less well understood than that of incremental improvement. Best practices, rules of thumb, and dictums like "Have lots of crazy ideas," or "Get creative, folks!" are par for the course for trying to incite innovative work. These may work some of the time, but do not provide any understanding as to how to "get creative".

The object of our work is to understand the dynamics of radical breaks in the redesign process, particularly in the context of designers working in small, horizontally organized teams. It is our hope that insights gleaned from this chapter will be applied to a broader range of design contexts.

1.4   Research Question

This study was designed to observe how small design teams make radical breaks. The data set I created offered more than just radical breaks, it offered numerous examples of incremental and mid-level changes to the engineering drawing each group got. The teams who made incremental improvements behaved very differently from those who made radical breaks, and by design the media they used was different.

The resulting formulation of the general research question is:

1.5   What Are the Critical Behaviors That Influence Outcomes?

This research suggests that how subjects determined what they would do and when subjects determined what they would do had a critical impact on outcomes. I have called the mechanism of determining how and when subjects determine what they would do "Path Determination". Under this rubric, I have uncovered two modes of behavior: wayfinding and navigation.

Wayfinding involves in situ determination of path based on perceptual cues from the real or imagined environment, often accompanied by narrative. Radical Breakers enlist wayfinding to discover, explore, and unpack novel product concepts and use scenarios.

Navigation involves a predetermined plan of what and how to redesign. Many of the tools and practices of optimization tacitly or explicitly rely on navigation to structure the activity of design. Groups who excel at incremental redesign engage in navigation behaviors.

2   Methodology: Case Creation for Building Frameworks

In the methodology developed for this study, I generally follow both Eisenhardt's (Eisenhardt and Graebner 2007) and Yin's (Yin 1994) work to build theory from case study research, with two significant modifications. The first modification is that I did not go into the field and collect data. Instead, following an approach suggested in a paper by Malte Jung et al. (Jung 2010), the cases I created are from a redesign task done in the lab, and not the result of field studies. The second modification concerns the claim I am making about the research. Unlike Eisenhardt and Yin, I do not make claim to building a theory per se, though I do claim to be presenting a descriptive framework.

The case study approach based on field data as compared with case creation based on laboratory work has advantages and disadvantages. One disadvantage to the case creation approach lies in the fact that the data are not collected in the wild. The activities of designers situated in their studios or industrial settings are extremely complex. Stakeholder demands, deadlines, team and management structure all have impact on how and what designers do. These factors are greatly truncated in a laboratory-based study. Design projects in the wild often take months. The temporal scope of this study was 30 min plus time for an interview.

Nonetheless, advantages of a laboratory study are many. The researcher can control many factors, which are out of control in the wild. Prompts can be supplied, teams can be handpicked based on numerous criteria, duration of observation can be calibrated, and media and tools that designers use can be filtered.

In this study, I designed an open prompt (which will be discussed presently); I strove to select teams that knew one another and had worked together; the relative brevity of the redesign task, 30 min, allowed me to look with fine granularity at the data that likely would have not been possible with notes made from field observations. In this study I attempted to control the media designers could use, sometimes with less success than I had anticipated. For example, several participants used their cell phones, which augmented the media I supplied to them.

Perhaps the greatest advantage of the case creation approach over case study is that the former affords iterative viewing and analyses. Often in studies done in the field, researchers are not permitted to video record proceedings for privacy and intellectual property reasons. This means that the researcher must rely on carefully-honed observation skills to filter real time data into notes. This is a very difficult procedure for even the most experienced researcher. Biases easily creep into the note taking, as there is only so much a person can take in, select, and note. In case creation approach is taken, cameras and microphones can be placed in multiple locations. Our study employed four cameras and five microphones in order to capture several angles including a top-down angle which allowed us to see precisely which markings were made on paper and when.

While I believe the insights gleaned from the study merited putting the designers into the lab, the applicability of these insights still need to be brought into a bigger arena of practice and tested. In order for the frameworks, which come from these studies, to be truly robust, there is no substitute for in vivo testing.

2.1 Data Collection: Video and Transcripts

Here I follow much of the work that has been done at Stanford's Center for Design Research starting with the construction of the Design Observatory by John Tang (Tang and Leifer 1988) and continuing with Scott Minneman (Minneman et al. 1995), Ozgur Eris (Eris 2002), Neeraj Sonalkar (Törlind 2009), and Malte Jung (Jung 2007). In these studies, participants at work on design and redesign tasks are recorded on several streams of video. As previously mentioned, in my study I used four streams of video: one for each of the three participants and one top-down camera above the table.

These video recordings are then made into transcriptions for careful evaluation and notation, as video can be difficult to notate.

2.2 Qualitative Analysis and Emergent Coding Scheme

In this study, I first present qualitative observations, showing examples of group generated media (sketches, lists, 3-D models), videos and still frames from video, and transcripts. I describe what I see and why it is important.

In order to insure these are not unique events, I count the occurrences of the qualitative events and compare the totals across teams. The totals reflect both the occurrences of the events and the total time of the events for each team.

I follow Jordan and Henderson (Jordan and Henderson 1995) in their approach to building a coding scheme without preconceived coding and involving fellow researchers in uncovering patterns and insights in the data, to the extent that is reasonable. However, in all honesty, I believe it is virtually impossible to formulate a coding scheme without some preconceived notions to start with. It could be argued that the data must speak for itself. In my experience, the data eventually speaks for itself, loud and clear. Nonetheless, the fact that a researcher is interested in a particular phenomenon means that he or she has preconceived notions. The trick is to get close enough to the data for a long enough period of time, so that your notions begin to become visible to you and are able change.

One of the advantages to taking a case creation approach, in comparison to hypothesis testing, is that you are not trying to prove a point. Instead, you are trying to unpack a complex group of phenomena, to see if you can detect patterns, and to see things you have never consciously seen before.

What is my point of view? I am a professional designer and design educator. I have a stake in finding "a better way" to practice and teach design. I came to this study with an explicit interest in how the shared media designers enlist shapes their conversations, and hence the development of new products. I found that media was only a part of the story, albeit an important one. The meaning of media is intimately connected to how it is used, in other words the behaviors that surround the media.

Through repeated viewing of teams at work, I was able to detect that teams made different kinds of "moves" which did the work of transforming concepts in different ways.

I have not applied statistical methods to the data as an analytic method. I believe that the data set I have created does not support statistical analysis. While I have counted events and noted duration of events, this does not constitute using mathematical modeling to make meaning of the data.

In this and other studies, I look for evidence of specific behaviors. In the course of these studies, I have had to discard several iterations of coding schemes along the way as new insights developed from examining the data. Furthermore, while I have worked closely with other researchers, I have often worked alone on both developing a coding scheme and coding data. One limitation of a 4-year study is having different researchers work with me over an extended time. I have taken care, however, to frequently consult with other researchers to make sure codes were clear and clearly applied to data.

I follow Bakeman and Deckner (Bakeman et al. 2008) in an attempt to make the codes mutually exclusive and exhaustive. This took some time, but was worth the effort. If a statement or behavior could not clearly be described by a code it was left out. While coding we would constantly refer to "the rules" of a given scheme. I discovered that it was really easy to forget what I was coding for at any given moment. Dr. Martin Steinert of Stanford University suggested a simple, but effective remedy which kept me on track: write "the rules" of the individual code on a slip of brightly colored paper and keep it front and center.

Creating "the rules" was a time-consuming process, but one that sharpened our eyes and brought greater clarity and confidence to the work. Sometimes I would have a specific, well-defined behavior I would set as a "rule". Most often I would have a general notion of how teams were behaving, something several groups would do, or something consistently occurring in some groups but not others. It took lots of enlightened trial and error to actually define what I believed I was seeing.

Even more difficult was translating and applying a high-level notion like "wayfinding" to a specific set of behaviors: there were days of wanting to drop the whole thing when I couldn't discern concrete examples of what I believed I saw, followed by days of clarity when I thought that a robust "rule" had crystalized. This was followed by days where the data didn't seem to support the "rule". Sometimes the right formulation would arrive in the shower or on long walks.

I have not applied statistical methods to the data as an analytic method. I believe that the data set I have created does not support statistical analysis. While I have counted events and noted duration of events, this does not constitute using mathematical modeling to make meaning of the data.

I present a framework for understanding the interaction of media and behaviors in small teams engaged in a redesign task. I make no claims to presenting hypotheses, let alone testing hypotheses. In this respect, the work of this analysis presents qualitative assessment in support of qualitative claims.

2.3 Participants: Non-homogeneous

Participants included a wide range of design engineers in the domains of product design and mechanical engineering: undergraduates, graduate students, professional product designers, professional engineers, and educators in both product design and engineering design. Specifically, we engaged a total of 42 participants, 26 of whom were male and 16 of whom were female. Their ages ranged from 19 to 77 years old.

Participants were grouped into non-hierarchical teams of three. With one exception, participants on a team were familiar with one another, either having worked or socialized with each other on many occasions. Our intent with putting together teams was to attempt to ensure that the participants would be comfortable with each other during the redesign task.

I have chosen not to overemphasize the participant in this study. This is because I am concerned with the subjects' behavior and how it influences outcome, rather than wishing to establish causal relations between the subjects, their backgrounds, or thinking styles and outcomes. While this is certainly a meaningful and appropriate undertaking, it is well outside of my expertise. I am a "student" of phenomenology and behavioral empiricism, thus I am mainly concerned with what subjects say, do and make.

2.4 Data Set: Triggering Radical Breaks

Several factors went into developing the data set. I wanted to do a close analysis of how designers perform radical breaks. Video recording of small teams at work promised to be a good approach, though the problem of how to capture radical breaks meant we had to set up proper conditions for them to occur.

Our notion of radical break entails making a significant departure from an existing solution. I came to see that "significant" meant a change in the core functionality of an object in the context of a new system, though at the time of constructing the redesign task I did not have a definition this specific in mind. Instead, I was simply looking for "big changes".

We created this data set in order to examine the phenomena with a mixed methods approach. We video recorded 14 teams in a 30-min redesign task, constructed to tease out radical breaks. We fashioned the redesign task and stimuli in order to create cases primarily for qualitative case study observation, rather than to focus on quantitative analysis, as the object of this study was to develop a framework about how radical breaks are accomplished, rather than perform hypothesis testing. We believe that qualitative insights harvested from our data will better serve designers wanting to make sense of their design processes than quantitative data attempting to prove effectiveness of a given approach.

2.5 Prompt

The prompt we designed stated:

> The object in front of you allows the identification of the properties of material objects. Redesign it.

Because we were interested in observing the effect of the media we provided, we decided to keep the prompt as brief and open as possible. The notion of "redesign" would have to be determined by each team.

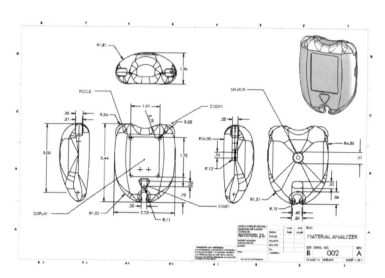

Fig. 1 Primary media, material analyzer

2.6 Media

All subjects were provided with a printed prompt and toleranced, labeled engineering drawing of a device that could reportedly analyze the properties of materials (0) that we termed "primary media". In addition, they were given a text based prompt. The two control groups were given the engineering drawing alone. The 12 other groups were given an additional model, which we called "secondary media" after 5 min.

2.6.1 Primary Media

The engineering drawing (0) offered six views of the object: front, back, two sides, top, and an orthographic view. The former five views were wire frames, while the sixth, orthographic view was generically surfaced. The major features of the device were labeled with call outs (e.g. "display", "focus", "zoom", "sensor", and "start"). While the engineering drawing was labeled "Material Analyzer", no description of how it worked was offered (Fig. 1).

2.6.2 Secondary Media

These models included a well-rendered foam model, a low-resolution ID-style sketch, a rough cardboard puck with features drawn on it with a marker, and a rough cardboard experience prototype (Fig. 2).

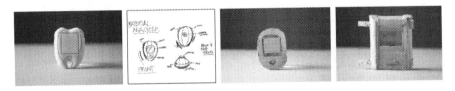

Fig. 2 Secondary media

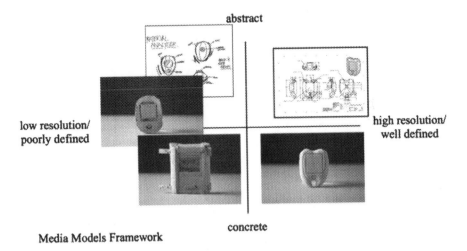

Media Models Framework

Fig. 3 Media models framework

Secondary media were chosen through use of the Media-Models framework (Edelman 2009; Grosskopf et al. 2010; Edelman and Currano 2011). The reasoning behind giving a diverse selection of media was to see how participants responded to media of different quality, as well as to uncover which kinds of media might induce a radical break.

The Media-Models framework was developed in order to understand the kinds of media used in diverse design environments. To briefly review, the Media-Models framework describes characteristics of the media that designers use in various stages of development. The framework utilizes three dimensions: abstraction, definition (resolution), and mutability (Fig. 3). Development stage-appropriate media can be enlisted for development based on these dimensions.

2.7 Data Sample

The total raw data for the 14 groups totaled 14 h of video, 11 three dimensional models, 74 pages of sketches, lists, notes, and 1,047 pages of transcripts. For each of the groups, four streams of video were combined into a single video file using Apple Final Cut Pro, a professional video-editing tool.

2.8 Coding Schemes

The following can sum up the subsets of the data I examined:

What subjects wrote and sketched
When subjects wrote and sketched
What subjects said
When subjects said it
How subjects gestured
When subjects gestured

2.9 Case Selection

Assessment of each team's deliverable, the redesigned object, was made through analysis of video streams and through participants' responses to interview question Q1, "How did you redesign the product, what does your new model look like, and how does it work?"

Each piece of data was watched, read, examined, handled and observations were recorded in notebooks. It became clear that while desirable, it would take an extremely long time to do a detailed analysis of all 14 cases. We needed to develop a way of choosing a limited number of cases. Eisenhardt suggests that the number of case studies should be between four and ten (Eisenhardt and Graebner 2007).

I chose forced ranking as a method and got four experts to do the ranking. The experts were all professional designers who also teach design at Stanford.

2.10 Results of Forced Ranking

Results of the forced ranking were combined. From these I chose four groups for close study:

Group 2: the least change and lowest relevance (to whom I refer as "Optimizers")
Group 10: the most change and highest relevance (to whom I refer as "Breakers")
Group 9: the least change with high relevance (to whom I refer as "User-Centered")
Group 7: the most change with low relevance (to whom I refer as "Wild Ideas")

Figure 4 presents the forced ranking results for all the teams. Groups 2, 7, 9 and 10 are circled. The media associated with each group is to the right of the chart. The reader may note that all three groups (Groups 2, 6, and 9) who ranked with lowest concept change all received the well-resolved foam model (highlighted in red), while the two teams associated with greatest concept change (Groups 7 and 10) received the cardboard puck (highlighted in tan) and the experience-like model (highlighted in blue) – both ill-defined models. Nonetheless, the rankings make it

Forced Ranking

Fig. 4 Forced ranking results for all fourteen groups

clear that the media given to the teams only tells part of the story. Three of the teams (Groups 3, 4, and 8) who received the cardboard puck did not make large changes to the concept, and one of the teams (Group 11) who received the experience-like model made mid-level changes to the Material Analyzer.

Figure 5 represents the four cases selected by forced ranking on a two by two, using the final graphic output of Groups 2, 7, 9, and 10. The y-axis denoting relevance and the x-axis denoting change of concept. Thus, the rankings here can be considered (clockwise, from the lower left corner): low change and low relevance (Group 2, "Optimizers"), low change with high relevance (Group 9, "User-Centered"), high change with high relevance (Group 10, "Breakers"), and high change with low relevance (Group 7, "Wild Ideas").

While I believe the forced ranking method was invaluable as a method for objectively selecting four teams for very detailed analysis, I recognize that the implementation may be flawed. The results seem to indicate a bias or confusion that conflates big changes with relevance. While this outcome is worthy of a study on its own, it was not what I was expecting. Perhaps the instructions to the experts were not clear; perhaps the community of design engineers at Stanford has a deep bias towards the new and different, as opposed to meaningful incremental change.

While the above considerations hold in theory, I believe in practice they do not have major implications for the work I have done. My primary concern was having an objective process for case selection. I did not want to be perceived as cherry-picking the cases in order to support my biases.

Through the initial detailed analysis of all 14 teams, I saw that all the teams exhibited similar behaviors depending on their deliverability. Some teams were more pronounced in what they did well and what they did poorly. If I had my druthers, I would have picked similar cases plus a case in which the team made mid-level changes to the concept presented in the engineering drawing.

Forced Ranking

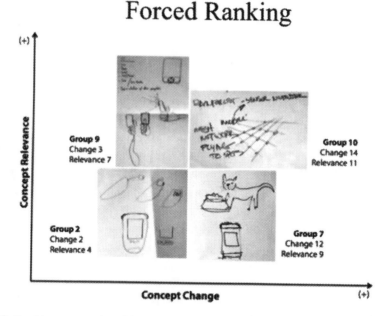

Fig. 5 Graphic representation of four groups selected for close study

2.10.1   Final Deliverables for the Four Cases

Below are descriptions of the four cases and a sample of their graphic output.

2.10.2   Group 2, "Optimizers"

Group 2, moved the sensor of the Material Analyzer so it would not be blocked by the user's hand while in use. They added a touch screen and put the sensor on a rotating device in order to make it easier to use (Fig. 6).

2.10.3   Group 7, "Wild Ideas"

Group 7, transformed the Material Analyzer into an additive for cat food that would keep cats away from people who are allergic to cats (Fig. 7).

2.10.4   Group 9, "User-Centered"

Group 9, made the Material Analyzer more ergonomically sound by extending the form to include a handle. This also solved the problem of the sensor being blocked

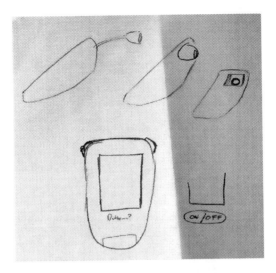

Fig. 6 Group 2 final concept sketch

Fig. 7 Group 7 final concept sketch

by the user's hand. They added some functionality by making suggestions for a graphical user interface, as well as changing the orientation of the buttons (Fig. 8).

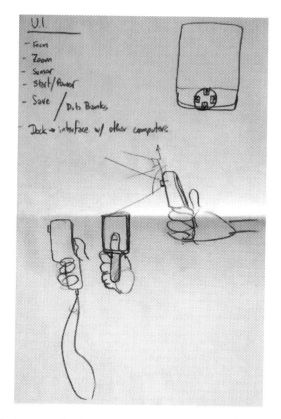

Fig. 8 Group 9 final concept sketch

2.10.5 Group 10, "Breakers"

Group 10, transformed the Material Analyzer into the "Mobile Mesh Network", a system of flying, solar powered robots, replete with on-board GPS, little drills to take core samples, conduct DNA analysis and wirelessly send the information back to a home base to be deployed in the Amazonian rain forest in search of new compounds for pharmaceutical research (Fig. 9).

3 Key Distinctions: Wayfinding and Navigation

Some subjects made plans of how they would spend the next 20 min, while others made no plans at all, and nonetheless had interesting outcomes. These observations developed into the key distinction "Path Determination" and were called "Navigation" and "Wayfinding".

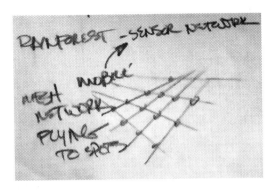

Fig. 9 Group 10 final concept sketch

My approach to observing and analyzing behaviors is based on contemporary work in sociology and behavioral science. I wanted to see if I could observe two classes of behavior, "wayfinding" and "navigation", in the actions of designers at work redesigning a handheld device, the Material Analyzer, in a laboratory setting.

"Wayfinding" and "navigation" are two methods of making our way through physical terrain proposed by contemporary social scientists. My interest in the nature of wayfinding and navigation is in the context of engineering design. The question I asked was "Do these behaviors determine or influence the outcomes of a redesign task?"

4 Literature: Anthropology and Archaeology

In his book "Lines: A Brief History," Tim Ingold (Ingold 2007) suggests two methods for making one's way in the world, wayfinding and navigation, which he illustrates with descriptions of how the Orochon (Kwon 1998) people hunt wild reindeer, riding saddled domesticated reindeer. In this passage, Ingold uses the word, "wayfaring" in place of "wayfinding" and "transport" in place of "navigation", though the sense of both is consistent with the use in this chapter.

> The path of the saddle-back rider, according to anthropologist Heonik Kwon, is 'visceral in shape, full of sharp turns and detours'. As they go on their way, hunters are ever attentive to the landscape that unfolds along the path, and to its living animal inhabitants. Here and there, animals may be killed. But every kill is left where it lies, to be retrieved later, while the path itself meanders on, eventually winding back at camp. When however the hunter subsequently goes to collect his kill, he drives his sledge directly to the site where the carcass has been cached. The sledge path, Kwon reports, 'is approximately a straight line, the shortest distance between the camp and the destination'. Not only is the sledge path

clearly distinguished from the saddle path: the two paths depart from opposite sides of the camp and never intersect. It is along the saddle path that life is lived: it has no beginning or ending but carries on indefinitely. This path is a line of wayfaring. The sledge path, by contrast, is a line of transport. It has a starting point and an end point, and connects the two. On the sledge the body of the dead animal is carried from one site, where it was killed, to another, where it will be distributed and consumed. (Ingold 2007)

Later, Ingold emphasizes that the activity of wayfinding is characterized by being perceptually attentive to the terrain as it appears in time:

... the traveller's movement — his orientation and pace — is continually responsive to his perceptual monitoring of the environment that is revealed along the way. He watches, listens and feels as he goes, his entire being alert to the countless cues that, at every moment, prompt the slightest adjustments to his bearing. (Ingold 2007)

Thus, wayfinding involves responding to immediate perceptual cues and information given by the environment. While the goal of the wayfinder may be predefined, the path is determined in the moment and in response to direct perceptual cues that are detected in the environment. Furthermore, while Orochon hunters do have a clear goal, it is actually their meandering path that is critical to reaching it. When Orochon hunters recount the hunt, they dwell on unusual things they saw or heard, not on the kill. The paths of wayfinders are often guided by narrative (Kwon 1998), and, like stories, they often meander, weave in and out, often turning back on themselves.

In contrast, the practices of navigation are linear. Orochon hunters wrap the kill, leave it in place, and head directly back to their village. They return post haste on sleds, in the most direct route possible, and bring the carcass to the village along the most efficient route.

While wayfinding can be described as determining one's path in response to direct contact with the environment, often accompanied by a narrative, navigation involves setting a predetermined route, the shortest possible path, often a straight line. Navigators rely on charts and maps, which provide a top-down view of all things at the same time. Gone are the narratives that play out in time. In their place are triangulation and charts, which reduce a myriad of perceptual input to thin lines on a grid (Webmoor 2005). While navigators do look at the landscape, it is always in reference to maps, charts, and instruments like compasses, which tell navigators where they are (Ingold 2000).

The practices of navigation, Ingold tells us, are well suited for keeping both the navigators and their payload intact and unchanged, not unlike the practices of optimization in engineering design. Wayfarers, on the other hand, experience change during the course of their journey and what they carry may change, not unlike what happens when engineering designers perform radical breaks.

The concepts of wayfinding and navigation can be put to work in understanding how design engineers work, and how they accomplish their redesigns. In the next section, I will give examples of how wayfinding and navigation can be seen in the data set we created.

5 Qualitative Evidence: Wayfinding and Navigation

Ingold's description of wayfaring and navigation are well suited for describing the behaviors observed in design engineers when they are engaged in redesign exercises.

5.1 Wayfinding: Feeling, Acting, and Narrating

Wayfinding refers to in situ determination of route based on perceptual cues often in the context of narrative. This was coded by identifying moments when changes to the product concept were made while subjects felt, acted, and narrated their way through real or imaginary environments. Rather than a top down approach of determining what is wrong with the product and following a program for fixing it, these designers feel, act, and narrate their way through, employing gesture to "feel" the object and the environment as the scenario unfolds.

Below (Table 1) is an excerpt from the transcript with an analysis of Group 10 exhibiting behavior I coded as wayfinding:

Nearly every move this group makes is in the context of feeling the real or imagined environment (Fig. 10). The concept of the glove co-develops with the gestures, and eventually becomes a bracelet with a thimble type sensor.

Group 10 made in situ determination of their path based on perceptual cues in their environment. By this I mean they built on one another's gestures and sketches.

Figure 11 above illustrates a condensed version of steps along the path that Group 10 (the Breakers) made. It is hard to imagine that this path was planned in advance.

I will describe the steps the group made, because I have found them extremely valuable in understanding how wayfinding is concretely achieved. Please note that the actual conversation contained many more suggestions and many more "moves". I have chosen these because Group 10 either wrote or sketched these moves. The significance of the "moves" is how they open up new possibilities for envisioning how the redesigned product could be. Furthermore, note well that the development occurs in the context of enactment and feeling in real and imaginary environments.

Below, in Table 2, I have paraphrased the transcript for clarity (though I believe I have been true to the conversation) and have provided an analysis of the wayfinding moves that Group 10, the Breakers, made.

The preceding transcript and analysis serves as an illustration of the kinds of moves that constitute wayfinding. It is not meant to be a complete list of wayfinding moves, but a start. These in situ determined moves stand in contrast to the moves of navigation, in which front-loading the plan of action is the primary strategy for moving through the problem space.

Table 1 Wayfinding

Transcript	Analysis
B: Yes, how about some kind of a glove thing. Cause you want to touch a surface, right? You learn a lot by feeling ...	[Subject B reaches out and feels the surface of the table]
A: Oh yea, the feeling it	[Subject A reaches out and feels the surface of the table]
B: ... the friction	
M: Oh, oh, nice. Nice	
B: You got multi-sensors for friction for hardness, for whatever. Then you've got the big sensor that collects the optical data	[Subject B feels the surface of the table with individual fingers, subject A follows suit]
You just rub things ...	[Subject B waves hand in the air, "rubbing" imaginary surfaces]
... like flesh	[Subject B turns to subject A and rubs the flesh of subject A's arm]
	Here, feeling the flesh of Subject A's arm provides the ground for a new functionality for the material sensor, this is an example of wayfinding
A: What if you want to touch things that are far away or scary? Then you could have ... I'm looking at this and thinking if it's touching wireless then you can throw it over there. Like it's on a string	[Subject A makes a throwing motion with both hands, away from body]
	Here, the notion of disconnecting the sensor from the CPU occurs in the context of a dangerous situation where you would want to throw the sensor into a substance, this is an example of wayfinding
B: Yea	[Subject B copies the gesture with two hands, and then one hand. Subject B moves fingers in a carefully articulated fashion]
B: That also gives you gesture	Here, a new concept for device control arises in the context of the gesture, this is an example of wayfinding
M: Ooooh, I like that	
B: That gives control	[Subject B turns hand over and makes a new gesture]
B: Like you go, give me that. Give me that	[Subject B points with index finger]
M: So, what's cool is you can do a whole series of gestures	[Subject M points with all fingers and gestures with hand]
	Here the enactment of using the hand for control suggests "a whole series of gestures", this is an example of wayfinding
You can point	[Subject M points with index finger alone]
and then you have a display	[Subject M touches wrist and back of hand]
So, this would be your display here	[Subject M sketches the display on a glove]
	Here the form factor of the imaginary glove and the enactment of use is the ground for the placement of the display, this is an example of wayfinding
So, let's say display is here	

(continued)

Table 1 (continued)

Transcript	Analysis
A: Right	
M: Maybe a couple buttons	[Subject M touches wrist with index finger several times, indicating button use and placement]
Then you have the different fingers. I love the ability to touch or point	[Subject M sketches the glove]
Then this is ... you're using your finger gestures to define functionality	[Subject M touches table top with index finger]
which comes up on the display. So, glove ...	
B: Bracelet ...	
M: I love that. Oh, bracelet is interesting	[Subject M points to wrist and then follows hand to index finger]
M: What about a bracelet with just a few connectors so it doesn't give you a whole glove. Cool. And a little ring thing	[Subject M wraps his index finger around another finger, indicating a ring] Here the glove metaphor transforms into a bracelet and a ring, this is an example of wayfinding
So, then it almost becomes a jewelry thing	

Fig. 10 Group 10, feeling the environment

5.2 Navigation: Plans of Action and Hierarchical Lists

I observed navigation through two behaviors, plans of action and hierarchical lists.
Below is an excerpt from the transcript of Group 2, the Optimizers, that indicates a plan of action coded as navigation:

Z: I think it's more of like – like re-placement of the various buttons and sensors in a more logical way.

This statement provides a clear statement of intention, and assumes that the work of redesign is a matter of rearranging buttons and sensors, as compared to no statement of intention. This seems to have the effect of giving a clearly defined

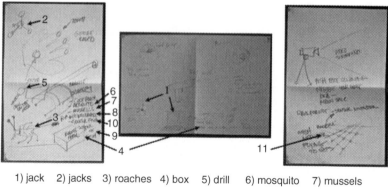

1) jack   2) jacks   3) roaches   4) box   5) drill   6) mosquito   7) mussels
8) hummingbirds   9) school   10) sucker fish   11) mobile mesh network

Fig. 11 Group 10, moves

target for the entirety of the redesign task, in comparison to responding to what appears on the perceptual horizon as a cue for what to do next.

5.2.1 Navigation: Hierarchical Lists

Significantly, groups who exhibit navigating behaviors often point to the engineering drawing in front of them or to features lists they generate. In these cases, the path of development is the result of creating a bullet point plan based on fixing what is wrong with the product in the engineering drawing. Figure 12 shows an example of a hierarchical list from Group 2:

Figure 13 is a photograph of Group 2 generating a features list. As previously mentioned, the list began to be generated very early in the session at 2 min and 52 s. One can see that apart from the features list, the table is clear of drawings and any other group-generated media.

5.2.2 Wayfinding: Non-hierarchical Lists

In contrast to the kind of hierarchical list making in which Group 2 engaged, Group 10 (the Breakers) made non-hierarchical lists. These were made as a record of concepts developed, rather than a roadmap or a series of to-be-accomplished bullets. At times, Group 10 used their list as a way of seeing where they had been in order to find new, unexpected connections.

Nearly all groups, with the exception of Group 10, displayed a mixture of both wayfinding and navigation behaviors. The behaviors of those groups who performed radical breaks were heavily weighted on the wayfinding side, while the behaviors of groups who optimized were heavily weighted on the navigation side.

Table 2 Wayfinding moves

Transcript	Analysis
A: What if it was a jack you can set down anywhere?	Step 1: "a jack you can set down anywhere", Subject A has just suggested a scenario in which you wouldn't want to touch the material to be analyzed. Subject A makes a throwing motion
	Move: Disaggregation. The sensor is separated from the material analyzer, and treated as a unique object. This can be considered as an instance of "breaking apart" an object to change it
B: It's a pocket full of jacks you can throw into a room	Step 2: "a pocket full of jacks you can throw into a room", Subject B expands on Subject A's throwing gesture, as well as suggesting a variation of how the user would interact with the sensor
	Move: Multitude. The sensor is now considered as a multitude of sensors
They are cockroaches that run around the room	Step 3: "they are cock roaches that run around the room"
	Move: Animate. Now the sensors are semi-autonomous, or at least self-moving
M: They live in a little box that gives you all the information	Step 4: "they live in a little box that gives you all the information"
	Move: Component. The sensor-roaches are reintegrated with the CPU part of the Material Analyzer
B: They have a drill on the nose, like a mosquito, they drill into surfaces	Step 5: "they have a drill on the nose, like a mosquito, they drill into surfaces", Subject B makes a drilling motion with his index finger pointing down at the table. This notion is introduced in the context of roaches getting information about the room. Some insects (though not roaches) do employ a proboscis to penetrate surfaces. The table is right in front of the participants, and they choose to integrate it into the scenario they are building. The table is constructed from a laminate, and Subject B suggests that the little drill could take core samples to analyze laminates
	Move: Add new function. Biomemesis, functional biology
M: They could be mosquitos that fly	Step 6: "mosquitos that fly", Subject M picks up on the mosquito reference and develops it, saying there could be different versions
	Move: Amplification. Biomemesis, functional biology
B: Other animals? Mussels that cling to the side of buildings	Step 7: "other animals? Mussels that cling to the side of buildings", Subject B asks about other animals and how they behave. Notice that the suggestion is not simply a noun, the suggestion is a noun a verb and an object; the formula here is a something (subject) that does something (a verb) to something (an object). This indicates that the exploration is not simply one of form, but of how and in what

(continued)

Table 2 (continued)

Transcript	Analysis
	environment the biological form operates. Cues are taken from this and transformed. For example, mussels do not normally cling to buildings, as buildings are above ground, and mussels beneath the sea. In this case, the mussel is an active metaphor that is transplanted into the context of the conversation (examining the sides of buildings) they just had
	Move: Biomemesis, functional biology
They could be hummingbirds that hover	Step 8: "hummingbirds that hover"
	Move: Biomemesis, functional biology
M: They could be a school of fish	Step 9: "a school of fish", this suggestion comes in the context of considering the Gulf Disaster, where you would need a water-borne device or devices. The notion of multitude is now considered as a coordination of elements, and thus a system
	Move: System
B: A suckerfish that suckles the bottom	Step 10: "suckerfish that suckles the bottom", this concept arises in the context of schools of fish, perhaps to address a tacit question, what kind of fish?
	Move: Biomemesis, functional biology
B: It is a mobile mesh network	Step 11: "a mobile mesh network", having considered a range of different kinds of multitudes and functions, Group 10 considers the implications of what they have explored. They have explicitly been considering each suggestion as a metaphor, and now they pull back from the metaphor to uncover the kernel that can represent all of these metaphors. This is Functional Abstraction. In this case, it is a network of mobile sensors that coordinate to map and sample different materials from different environments
	Move: Functional Abstraction

5.3 Codable Indicators: Wayfinding and Navigation

Path Determination (wayfinding and navigation) is defined as:

How and when subjects determined what they would do.

I looked at the conversations and the media generated by the groups for indications of wayfinding and navigation. Below are the four code prompts I used to analyze the data:

Navigation: plans of action
Wayfinding: in situ determination of next move
Navigation: hierarchical lists
Wayfinding: non-hierarchical lists

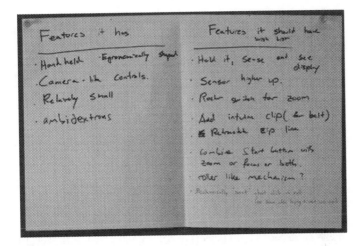

Fig. 12 Hierarchical features list

Fig. 13 Generating a hierarchical list

The first of these codes, plans of action, makes note of statements subjects made indicating a predetermined route.

The second code, in situ determination of next move, makes note of when a group changes course as a result of subjects feeling, acting, and narrating their way through real or imagined environments.

The third and fourth codes, hierarchical and non-hierarchical lists notes the characteristics of lists generated by subjects.

Table 3 Total hierarchical lists

Group 2 (optimizers)	2
Group 7 (wild ideas)	1
Group 9 (user-centered)	3
Group 10 (breakers)	0

6 Descriptive Results for Path Determination

In this section we will review the summation of events for Groups 2, 7, 9, and 10 in respect to Wayfinding and Navigation.

6.1 Navigation

Hierarchical lists were coded as indicators of navigating behavior. Below (Table 3) is a chart indicating the number of hierarchical lists made by the four groups we examined.

As you can see, Groups 2, 7, and 9 created 2, 1, and 3 hierarchical lists, respectively, while Group 10 made no hierarchical lists.

6.2 Wayfinding: Can It Be Measured?

Because wayfinding is central to our understanding of the mechanics of radical breaks, it is important to consider the ways in which wayfinding can be measured. As a first step, let us consider our working definition of wayfinding.

Wayfinding is in situ determination of path (in this case unpremeditated changes to product concept) based on perceptual cues in the real or imaginary environment (in this case indicated by a gestural enactment).

In order for us to change this working definition into a means of measuring, we needed to find and make note of each new product concept or change to the Material Analyzer, as well as record indicators of perceptual cues in the real or imaginary environment. While we have not tracked singular events in which participants made choices about where to proceed in the context of feeling through an imaginary or real environment, we have counted total changes to the Material Analyzer and compared them to total gestural enactments combined with narrating while sketching.

6.2.1 Unique Ideas

We counted all verbally expressed changes made to the original Material Analyzer. We then combined identical changes and tallied them as a single concept. This collection of single, unique concepts we call "major ideas". Table 4 represents the major ideas of each of the four groups.

Table 4 Major ideas

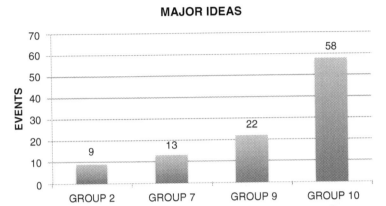

6.2.2 Gestural Enactment

We define "gestural enactment" as situated gestures with and without stimuli, in other words, "acting out". Enactments occurred when participants gesticulate in concert with exploring what the object is or could be and where it could be used, literally acting out the place and use of the object. Gestures included pointing, when the pointing was in the context of an enactment, though not in the context of simply pointing at a drawing, an object, or a fellow participant. Our notion of gesture did not include hand waving while trying to work out an idea, unless it was clearly in the context of acting out the use of the object, like stirring a cup.

Gestures created three codes for the size of gestures:

Level 1: from the wrist to the fingers
Level 2: from the elbow to the fingers, in relation to the table
Level 3: full arm away from the table

Bekker (Bekker et al. 1995) and McNeil (McNeill 1996) have observed several categories of gesture including Kinetic: "The movement executes all or part of an action performance" (Bekker et al. 1995). One kind of Kinetic gesture Bekker denotes as a walkthrough. Walkthroughs are "Sequences of kinetic gestures often used to describe the interaction between the user and a designed product" (Bekker et al. 1995). Thus we coded for the kind of gesture that Bekker considers a walkthrough, however we have departed from both McNeil and Bekker in that we have coded for finger gestures as well as hand and arm gestures.

Below is a chart (Table 5), which represents the total enacted gestures we counted for each of the four groups.

Table 5 Total gestural enactment

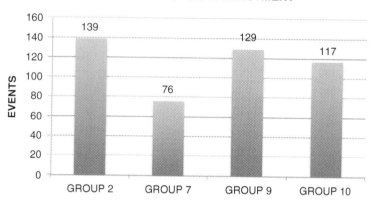

Table 6 Narrating while sketching

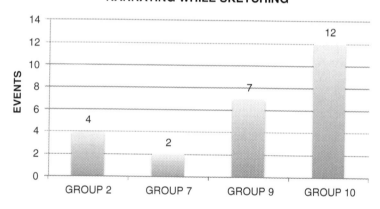

6.2.3 Narrating While Sketching

We noted that there were moments when participants narrated rich scenarios about a product concept and its use, but had little or no gesture at all. Sometimes these narratives were accompanied by drawings that depicted the user, the object, and the world, or parts thereof, thereby indicating Dimensions of Engagement. Frequently in these situations, the participant would mention new possibilities for use and product concept. We counted instances of narrating while sketching. Below is a chart (Table 6) that represents each group's observed instances of narrating while sketching.

While ideal, it is beyond the scope of this investigation to situate each unpremeditated change to the Material Analyzer in its original context to determine if it

Table 7 Total enactments plus narrating while sketching

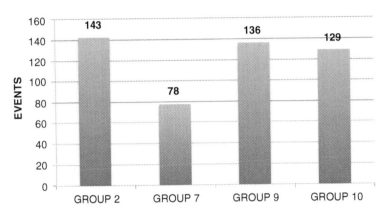

Table 8 Wayfinding score matrix

Group	Enactments + narrating while sketching		Ideas		Wayfinding
Group 2	9	÷	143	=	.06
Group 7	13	÷	78	=	.17
Group 9	22	÷	136	=	.16
Group 10	58	÷	129	=	.45

arose during an enactment. Thus, I suggest a proxy for wayfinding that involves dividing total unique ideas by total enactment events combined with narrating while sketching. Table 7 compares the unique ideas generated by each group as well as total enactments combined with narrating while sketching.

Again, the proxy I suggest for wayfinding involves dividing total unique ideas by total enactment events combined with narrating while sketching. Table 8 illustrates the results in matrix that illustrates the calculations for wayfinding:

These results can be easily visualized in Table 9:

6.2.4 Wayfinding with Major Ideas

As the reader can see, Group 10 had the highest wayfinding score, while Group 2 the lowest. Group 7 and Group 9 are very similar, and in the order they were ranked by experts for both change of concept and relevance of concept.

While not based on individual acts of change in response to enactments, this approach serves as a proxy for counting individual instances of enactment and resulting changes to the product concept. I offer this formulation as a first step in

Table 9 Wayfinding major ideas

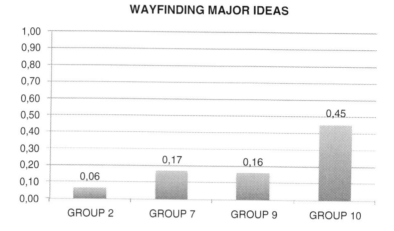

attempting to characterize and measure wayfinding in the domain of engineering design, and not a final and tested hypothesis.

7   Discussion: Wayfinding and Navigation

In the following sections I will consider each team in respect to wayfinding and navigation.

7.1   Group 2 (Optimizers): Navigators

Group 2's statements of plan of action and use of hierarchical lists, in combination with a low wayfinding score, suggest that they have a strong bias towards navigating behaviors.

Successful navigation, it seems, is supported by clearly stated plans of action made early on and the use of hierarchical lists.

7.2   Group 7 (Wild Ideas): Wanderers

This group had one hierarchical list, though in practice it is not clear how much they stuck to the plan afforded by the list. While they jumped from product concept to product concept, there was little enactment associated with their idea generation. They have a mid-level wayfinding score of .17. Because of these factors I suggest that they have a moderately low bias towards wayfinding, though it is not conclusive. This group seems to have engaged in wandering, but not wayfinding. Simply "wandering", moving from unrelated idea to unrelated idea is not wayfinding; wandering is analogous to making associative lists, in that both wandering and associative lists require little or no paying attention to a real or imagined perceptual horizon. This stands in contrast to wayfinding, which is a structured activity that involves using perceptual cues in the real or imagined environment to determine where to go next.

7.3 Group 9 (User-Centered): A Mix of Both

Group 9 lands right in the middle of the four groups. They generated three hierarchical lists to which they referred, and which structured much of their interaction during the redesign scenario. They generated 22 unique ideas and I recorded 136 instances of enactments combined with narrating while sketching. This led to a mid-level wayfinding score of .16, slightly lower than Group 9. I suggest that their behavior was a blend of wayfinding and navigation.

7.4 Group 10 (Breakers): Wayfinders

Group 10 had no instances of plans of action and did not generate a hierarchical list that explicitly structured their redesign work. I suggest that this indicated they did virtually no navigating. On the other hand, Group 10 has a wayfinding score of .45, the highest in the group, having generated 58 unique ideas during the course of 129 enactments combined with narrating while sketching. Because of these factors, I suggest that they have a strong bias toward wayfinding.

Based on the data from these cases, I suggest that successful wayfinding is achieved through a strategy of feeling, acting and narrating one's way through real or imagined environments. In addition, the cases suggest various tactics like transposition and simplification, remapping, disaggregation, creating multitudes, and systems in the context of the above strategy can be key to fruitful wayfinding.

8 Conclusion

Path Determination describes how designers use their tools to move through conceptual and perceptual landscapes. They can chose to enlist a set of behaviors that often lead to incremental improvements, or they can chose to enlist a set of behaviors that often lead to radical breaks. The former outcomes we have shown to be associated with navigation, the later outcomes we have shown to be associated with wayfinding.

One of the surprising implications of this study is that contemporary engineering practices tacitly enlist media and behaviors that encourage navigation. Hidden beneath the veil of mathematically-based scientific practices, the behaviors of navigation have remained unexamined because they are ubiquitous, and consequently unseen as behaviors in professional engineering and engineering education. What is striking is the revelation that engineering education in fact teaches navigation, while excluding wayfinding.

From an educational point of view, it may be that if you teach wayfinding, it is easier for engineers to be open to new and unexpected concepts, an important design behavior. This could be because in comparison to navigating, in which the target and path is clearly delineated, wayfinding asks us to pay close attention to what is new and unexpected on our perceptual horizons, rather than a map.

The lesson here is that wayfinding is not a primitive, intuitive skill that we just happen upon. Instead, wayfinding is a set of nuanced behaviors that require fine-tuning of perceptual abilities and unique tools, in the same way that navigation requires fine-tuned analytic abilities and unique tools. I have chosen to emphasize wayfinding over navigation not because it is more valuable, but because of its relation to radical breaks and the relative opacity wayfinding appears to have among design engineers.

Acknowledgments This research was made possible by a generous and on-going grant from the Hasso Plattner Institute for Design Thinking Research. The author would like to express his gratitude for their help, guidance, and patience without which this work would not have been possible:
Danke!
Jonathan Antonio Edelman
Stanford University
January 2012

References

Bakeman R, Deckner DF, Quera V (2008) Analysis of behavioral streams. In: Handbook of research methods in developmental science. Blackwell Publishing Ltd., Oxford, pp 394–420

Bekker MM, Olson JS, Olson GM (1995) Analysis of gestures in face-to-face design teams provides guidance for how to use groupware in design. In: Proceedings of the 1st conference on designing interactive systems: processes, practices, methods, and techniques, ACM, Ann Arbor, pp 157–166

Edelman J, Currano R (2011) Re-representation: affordances of shared models in team-based design. In: Design thinking, Understanding innovation. Springer, Berlin/Heidelberg, pp 61–79

Edelman JA, Leifer L (2009) Hidden in plain sight: affordances of shared models in team based design. In: Proceedings of the 17th international conference on engineering design (ICED'09), vol 2. ds publications, pp 395–406 (Print)

Eisenhardt KM, Graebner ME (2007) Theory building from cases: opportunities and challenges. Acad Manag J 50(1):25–32

Eris O (2002) Perceiving, comprehending, and measuring design activity through the questions asked while designing. Doctoral dissertation, Stanford University

Grosskopf A, Edelman J, Weske M (2010) Tangible business process modeling, Äì methodology and experiment design business process management workshops. In: Lecture notes in business information processing, vol 43. Springer, Berlin/Heidelberg, pp 489–500

Ingold T (2000) The perception of the environment: essays on livelihood, dwelling and skill. Routledge, London (Print)

Ingold T (2007) Lines: a brief history. Routledge, London (Print)

Jordan B, Henderson A (1995) Interaction analysis: foundations and practice. J Learn Sci 4(1):39–103 (Print)

Jung M (2010) Designing perception-action theories – theory-building for design practice. In: Proceedings of the eighth design thinking research symposium, Sydney (Print)

Jung MF, Ade M (2007) Design knowledge coaching – a conceptual framework to guide practise and research. In: 16th international conference on engineering design. ds publications. pp 415–416, (Print)

Kwon H (1998) The saddle and the sledge: hunting as comparative narrative in Siberia and beyond. J R Anthropol Inst 4(1):115–127 (Print)

Luebbe A, Weske M (2011) Bringing design thinking to business process modeling. In: Design thinking, Understanding innovation. Springer, Berlin/Heidelberg, pp 181–195

Luebbe A et al (2010) Design thinking implemented in software engineering tools. In: Staudinger I (ed) DAB documents, Sydney (Print)

McNeill D (1992) Hand and mind: what gestures reveal about thought. University of Chicago Press, Chicago (Print)

Meinel C, Leifer L, Plattner H (2011) Design thinking: understand – improve – apply. Springer, Berlin/Heidelberg (Print)

Minneman S et al (1995) A confederation of tools for capturing and accessing collaborative activity. In: Proceedings of the third ACM international conference on multimedia, ACM, San Francisco, pp 523–534

Tang JC, Leifer LJ (1988) A framework for understanding the workspace activity of design teams. In: Proceedings of the 1988 ACM conference on computer-supported cooperative work. ACM, Portland, pp 244–249

Törlind P, Sonalkar N (2009) Lessons learned and future challenges for design observatory research. In: 17th international conference on engineering design (ICED'09), Stanford University: ds publications (Print)

Webmoor T (2005) Mediational techniques and conceptual frameworks in archaeology. J Soc Archaeol 5(1):52–84

Yin RK (1994) Case study research: design and methods. Sage, Beverley Hills

Part III
Design Thinking Research in the Context of Embedded Business Teams

The Designer Identity, Identity Evolution, and Implications on Design Practice

Lei Liu and Pamela Hinds

Abstract This chapter describes the preliminary results of a study of designer identity, including what a designer identity is, how it evolves as a result of ongoing work-related interactions, and how it may influence design work practice. In our ethnographic research, we closely observed 12 in-house designers as they did their work in a major Chinese communication technology company. We found that designers identified with the design occupation in different yet non-mutual-exclusive ways, and that the way in which designers identified themselves influenced their creative thinking, brainstorming processes, and interactions with clients.

1 Introduction

Our work is motivated by two key limitations in existing research on design thinking and design practice. The first limitation is a lack of understanding about design work from designers' own points of view: How do designers think about themselves as a designer? This is a question about identity, one of the key foundational concepts employed to explain why people think about situations and others the way they do and why people do what they do in those situations (Ashforth et al. 2008: 334). A limited understanding of designer identity prevents us from accurate interpretations of design work. The abandonment of a particular design practice, for example, may not be due to efficiency or other instrumental calculations but really be a consequence of a perceived threat to the designer identity.

The second limitation concerns the complexity of work relationships involved in design work. Existing research on design work has primarily focused on the interaction among designers and between designers and users. Most of the time,

L. Liu • P. Hinds (✉)
Department of Management Science and Engineering, Stanford University, Stanford, CA 94305-4026, USA
e-mail: liulei@stanford.edu; phinds@stanford.edu

H. Plattner et al. (eds.), *Design Thinking Research*, Understanding Innovation, DOI 10.1007/978-3-642-31991-4_10, © Springer-Verlag Berlin Heidelberg 2012

however, designers in organizations often have to work with other stakeholders such as product managers, material specialists, modeling engineers, and clients, to name a few. To the extent that designers are successful at convincing others that their approach is valid, the designer identity is reinforced and the current design practices are sustained. To the extent that they are not, designers experience threats to their identity, which may in turn lead to modification of design practices.

In this research, we set out to understand and describe what it means to be a designer (i.e., the designer identity), how this may change as designers engage in ongoing interactions with other people at work, and how such identity dynamics may influence design work. To this end, we adopted a phenomenological approach and closely examined the "life-worlds" of a group of in-house industrial designers in a major Chinese communication technology company. We report some of the preliminary insights about designer identity generated from a careful documentation of the designers' day-to-day interactions with one another and with other people at work. Despite the fact that this research was done in one organization within a particular cultural context, we gain insight into a general way of studying designer identity that could be applied to other contexts defined by different national, industrial, and organizational characteristics.

2 Identity and Identity Work

Identity is one of the most popular topics in the social sciences. In our research, following the symbolic interactionist tradition, we define identity as a part of the self composed of the meanings that a person attaches to a particular role he or she typically plays in society (Stryker and Burke 2000: 284). A person can have multiple identities, as he/she is playing different roles in life, that is, as he is engaging in different types of social relations. A white male designer, for instance, has a racial identity, a gender identity, and an occupational identity; and each of these identities is associated with a particular situation that is defined by a set of social relationships. The designer identity, as an occupational identity, is constituted by a range of meanings designers attach to what they do at work and their work relationships with others. These meanings include, for example, task rituals, standards for proper and improper behavior, work codes surrounding relatively routine practices, and compelling accounts attesting to the logic and value of these rituals, standards, and codes (van Maanen and Barley 1984: 348).

The designer identity, like other identities, is fluid and mutable rather than static as designers' understandings about the designer work role may change over time. People make identity claims by conveying images that signal how they view themselves or hope to be viewed by others. By observing their own behavior as well as the reactions of others, who accept, reject, or modify these images, they maintain or revise their identity claims. Such a dynamic process is called identity work, which is typically defined as people's engagement in forming, claiming, maintaining, or revising their identities (Snow and Anderson 1987: 1348; Ibarra and

Barbulescu 2010: 137). Identity work is motivated by people's need for social validation and self-esteem, the former referring to people's desire to seek evaluations that confirm the meanings they attach to an identity and the latter to people's desire to view those meanings in positive terms (Ashforth and Kreiner 1999: 417; Swann 1987: 1038–1039). Identity work can be done by a variety of means including, for example, procurement or arrangement of physical settings and props, the arrangement of personal appearance, selective association with other individuals and groups, and verbal construction and assertion of identities (Snow and Anderson 1987).

In examining the related literature, we found no studies that explicitly examined designer identity and the relationship between designer identity and design work practice. The most relevant work is on design professions, which typically concerns specifying domain-specific bodies of knowledge in design professions (Carvalho et al. 2009) or about defining particular constructs (e.g., "creative act") that may be adopted to reach a unified understanding of different design professions (Wang and Ilhan 2009). These domain-specific knowledge and constructs are seen as the critical meanings that constitute the designer identity. Studies in this tradition are largely prescriptive and we lack an understanding about, first, how designers actually mobilize the different bodies of domain-specific knowledge to define a "self" in the workplace and, second, how the way in which the designer identity is constructed might influence the way in which design work is done. This study aims to fill these gaps.

3 Method

To explore questions about designer identity and its influence on design practice, we carried out a two-stage ethnographic study with a group of industrial designers. The first stage was conducted in the summer of 2010 and the second stage in the summer of 2011. Since we have not yet finished analyzing data collected from the second stage, here we only report the methods and findings of the first stage of our study.

This research was conducted in one of the in-house design centers in a major Chinese communication technology company engaging in the research, development, manufacturing, and sales of mobile phones and accessories. Although we started our field work with general observations of all three types of designers in the design center – industrial designers, user interface designers, and packaging designers – we later concentrated on the industrial designers whose work relationships and identity dynamics were particularly interesting. The findings reported here are, therefore, based on our analysis of the industrial designers' day-to-day work.

Design work was organized on a project basis. Once the design center received a design project from a product department, the design director allocated industrial designers to the project. Each designer assigned to the project developed one or more design proposals – which involved thinking, surfing on the Internet for inspirations/ideas, sketching, and documenting the design in a 2D file called CMF. Then all design proposals (from all those assigned designers) were pooled together for selection, first within the Center and then by the product department that initiated the project. Once one or more design proposals were selected, the designers who had come up those proposals moved on to complete the designs by working together with structural engineers, technique specialists, product management personnel, and others. Each industrial designer typically had one or two projects in hand at any one time. It usually took two to five work days for a designer to complete design proposals for a project, depending on the scope of the project, type of phone (easier candybar phones vs. more complex flip or slide phones), and the number of designs required. If a design was selected and eventually went in to production, then the designer's follow-up work lasted anywhere between 3 months to over a year, depending on the engineering and testing progress.

Data were collected using three primary methods: semi-structured interviews, observation, and archival analysis. We conducted 14 formal, face-to-face interviews with designers, which typically lasted 90 min, ranging from 75 min to 2 h. Although we prepared an interview protocol, our primary goal was to understand the designers, how they worked, and what it meant to be a designer from their perspective, so the interviews were driven by what they told us was important and interesting. The interview protocol also evolved over the course of the study as we learned more and identified avenues for fruitful exploration. In general, we asked questions about each respondent's view of design work and the designer profession, regular work-related activities, current and/or previous project(s), and relationships with other occupational groups within the company. All of the interviews were conducted in Chinese and tape-recorded with consent. The interviews were later transcribed and translated into English, yielding a corpus of 421 single-spaced pages.

Whereas interview data offer evidence of the content (i.e., specific meanings) of designer identity, data collected by observing designers at work allowed the discovery of how particular meanings are made from and/or manifested in social interactions. Observation is even more crucial for tracing changes of the designer identity over time. During the 4 weeks of our field work, on average we observed people's work at the Center 5 days a week, 6–7 h a day. Our typical routine was to arrive on site at 9 a.m., accompany one or more designers as they went about their day's work until 5 p.m., go out to have dinner (sometimes with members of the Center), and return to the office to complete that day's field notes. When we were observing, we typically carried a letter-size notepad, an audio-recorder, and a video-camera. Audio and video recordings were done with permission; and the recorded conversations, images, and videos were reviewed at the end of the day to fill gaps in the field notes. On average, we compiled 14 pages of single-spaced field notes each day, yielding a total corpus of 209 pages. We also recorded a total of 152 images and 71 video clips.

Finally, we collected work-related documents including, for example, annual reports (or "Results Announcements" 2006–2010), governance and organizational charts, the design center's roster, product catalogs (2008–2010), weekly work schedules, and project management documents. These documents not only helped us to understand the organizational context in which design work was done but also corroborated our interview and observation data. Project management documents and weekly work schedules, for example, situated what we had observed in a project context so that we were able to identify with whom a particular designer was likely to interact during a particular period of time.

Our data analysis was conducted following a set of guidelines recommended by scholars specialized in ethnographic research and grounded theory building (Strauss and Corbin 1998; Charmaz 2006), and using NVivo qualitative analysis software. Overall, it is an iterative process of examination and coding of data collected from the field, guided by a set of coding procedures recommended by Charmaz (2006).

4 Insights

Our preliminary research has revealed several insights related to designer identity, including what we call the multiplexity of designer identity, the evolution of designer identity, and identity's influences on design work.

4.1 Multiplexity of Designer Identity

We found that industrial designers mobilized three distinct, more or less mutually antagonistic meaning systems – that of art, engineering, and business – to construct the designer identity. In accordance with existing research on occupational identity, we refer to each of these meaning systems as an occupational rhetoric (Fine 1996) – rhetoric of art, rhetoric of engineering, and rhetoric of business. These three dominant and distinct occupational rhetorics of designers were identified from a close examination of hundreds of incidences where the industrial designers we studied described "who they are", "what they do" and "how they do what they do" in day-to-day work.

Each rhetoric consisted of a coherent set of concepts and expressions with embedded values which, together, constituted a way of talking about the substance and purpose of design work, thus framing the way people understand and act with respect to designers and their work.

Also, the use of a particular rhetoric was not exclusive to particular designers. Rather, all three rhetorics appeared in each and every designer's accounts for work, albeit designers differed in the relative emphasis on those rhetorics. Therefore, the designer identity, examined either at a group level or at an individual level, did not reflect a single coherent frame but rather multiple rhetorical stances and involved the maneuver, shift, and negotiation within and across these distinct rhetorical stances. In this sense, the designer identity was multiplex. Table 1 illustrates such multiplexity.

Table 1 Summary of designers' occupational rhetorics

	Rhetoric of art	Rhetoric of engineering	Rhetoric of business
Image (Ideal-typical social image)	Artist	Engineer	Business-man/woman
Substance (Key concepts in design work)	Appearance, style, image, spirit, soul, rhythm, product moral (or allegory)	Structure, function, material, space, fit, (processing/machining) technique, standard	Cost, vol. production rate, brand, branding strategy, product line, market niche
Value (Fundamental purpose of design work)	Beauty Novelty (uniqueness)	Usefulness Usability Feasibility Reliability	Cost-effectiveness Consistency Sustainability
Organization (How design work should be enacted)	Either "leave us alone" or "make us central"	Partnership with structural engineers and technique specialists	Superordinating product management teams

In Table 1, designers' three occupational rhetorics are compared and contrasted by the four major aspects of a meaning system – image, substance, value, and organization of work – which were inductively derived from the data. *Image* refers to the widely recognized, ideal-typical social image workers evoke to describe themselves and their work. In our research, when designers were describing who they are and what they do, they sometimes made analogies to other occupations, therefore invoking images of artist, engineer, and business-man/woman.

Substance of work refers to what designers see as key concepts in their work. A concept is a word or a phrase that expresses an abstraction formed by generalization from particulars. The concept of "style" in design work, for example, is abstracted from a wide range of specific styles and refers to a distinguishable set of forms – forms of color, shape, texture, or their combinations – that occur repetitiously in a number of products. Specific styles frequently mentioned by designers included, for example, minimalist style (featured by basic and a small number of geometric shapes, natural textures and colors, and clean and fine finishes), organic style (featured by curves, smoothness, and streamlines), classical Chinese style (featured by traditional Chinese forms such as the patterned color and shape combinations in palace adornment), and so on. As we can see in the table, the three sets of concepts are essentially different, corresponding respectively to the more sensuous aspects, the more tangible aspects, and the cost-effective aspects of the object of design work.

Value of work refers to the positive quality that renders a particular occupation's work desirable, either to the broader society or to particular communities, organizations, or individuals. In our data, the value of design work primarily existed in designers' statements about the purpose of their work, or answers to questions about what industrial design is for. A wide range of purpose statements were extracted, from broader ones like "to make cool products" and "to create a better life," to more specific ones such as "to make the product look good," "to make handy and tough cellphones," "to find out problems ... and solve them," "to define a kind of temperament for our brand and products," and "to make a living." From all these purpose statements, three distinct sets of values (of design work) were identified: aesthetic values that satisfy people's need for beautiful and/or unique appearances of products, engineering values that satisfy people's need for smooth and/or reliable functions of products, and business values that satisfy the company's needs for profit, brand development, and sustained growth.

Finally, *organization* of work refers to the way in which a particular occupation's work is integrated into the broader frame of work in a multi-occupation work environment. Specific to this study, it refers to the way in which designers think their work should be enacted within the company. Of primary concern here is the relationship between designers and other occupational groups in product development work. For example, whereas a hierarchical relationship entailed control and hand-overs of work, a partnership relation implied interdependence and teamwork. By framing their work in a particular way, designers explicitly or implicitly suggested some structure that was "right" for product development within the company. Framing design work using a rhetoric of art promotes "freedom" of design work, which can be achieved by either keeping designers independent from other occupational groups or placing them at the center to leverage (rather than be leveraged by) other areas of expertise. A rhetoric of engineering highlights the interdependence between designers and structural and material engineers, despite the fact that designers actually lacked negotiation leverage in the frequent conflicts with the engineers. A business framing of design work endorses designers' jurisdiction over product management teams by identifying design as the essential driver for successful product and brand development.

4.2 Evolution of Designer Identity

The three alternative rhetorical stances functioned as discursive resources which designers drew on in different ways in different situations to achieve their particular purposes – be it a specific interest-based purpose (e.g., "I need to argue with the product manager in this way to keep my design intact") or a general reflective purpose to make sense of what it is to be an industrial designer (e.g., "Am I the one who creates new styles or the one who solves existing problems?"). Depending on how others responded to the different framings, designers adjusted their strategies of rhetoric use – for example, switching from one rhetoric to another in a particular situation. It was such selective use of and strategic adjustment to alternative rhetorics that drove the development and change of the designer identity.

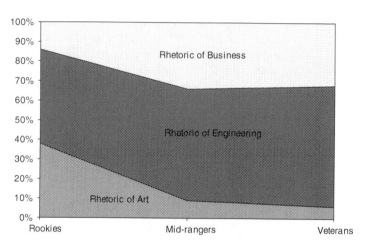

Fig. 1 Relative frequency of rhetoric use by tenure stage

Identity change can be said to happen when, for example, a particular rhetoric becomes significantly more popular while another one loses favor. Changes in the content of one or more rhetorics – that is, the concepts and values that constitute a particular meaning system – also represent a form of identity change. The ideal way to examine identity evolution is to track these changes over a long time frame, say, years or decades. Although the time frame of our first stage of field work is too short to track the longitudinal evolution of designer identity, one way in which the data allow us to gauge identity evolution is to compare rhetoric use across designers at different tenure stages. The assumption is that, if changes over time in a person's occupational identity are a consequence of long-term involvement in work, then systematic differences in identity should exist between those with longer tenures and those with shorter tenures.

In our observation, while all three rhetorical stances were adopted by all designers, they were not evenly distributed. Whereas junior (in terms of tenure) designers seemed to use the art and engineering rhetorics more than the business rhetoric, senior designers seemed to rely more on the engineering and business rhetorics than on the art rhetoric. To quantify such differences, we compared the relative frequency of different rhetorics across tenure groups. Specifically, we divided the designers studied into three groups: rookies, mid-rangers, and veterans. Rookies were those with a tenure of 6 months or shorter; mid-rangers were those with a tenure between 6 months and 2.5 years; and veterans were those with a tenure of 2.5 years or longer.

For each tenure group, we counted all the incidences where a particular rhetoric was adopted by the group members to describe and/or to justify design work. We then calculated the within-group relative percentages of art, engineering, and business rhetoric instances. The resulting 9 percentages (3 relative percentages in each of the three tenure groups) produced a chart of relative frequency of rhetoric use by tenure stage, see Fig. 1. The chart illustrates that, as designers became more and more experienced in work, there was a general tendency to talk about work in more and more engineering and business terms than in artistic terms. The drop of the artistic rhetoric and the boost of business framings seemed to happen very fast, within the first 6 months of working in the company.

What was driving such identity evolution seemed to be some important organizational and institutional factors that forced designers' identity work toward particular directions. First of all, the way in which product development work was organized in the company placed designers in a vulnerable position, making it extremely difficult for them to leverage their unique artistic expertise in negotiations with other occupational groups. For rookies, artistic framings were important because they believed such framings were the only leverage they had in dealing with other occupational groups. However, as interactions proceeded, they became aware that design work was frequently dominated by those who had no design or art trainings – it was these people who made decisions about what to design and whether a design was good or bad. Having realized that negotiating using artistic framings was futile, designers began to resort to alternative rhetorics to achieve their ends (which were primarily about keeping a design alive or faithful to the designer's original intent). We therefore saw instances where designers used engineering rhetorics to deal with product managers, who were able to interfere with the artistic aspect of design work but not so much with the engineering aspect.

In addition, broader societal and industrial factors exercised influence on designers' identity development. For example, the modern Chinese society has prioritized pragmatic values (e.g., usefulness, durability, and cost effectiveness) over idealistic ones. Reflected on product design, the artistic aspect was frequently said to be "secondary" or "non-substantial" compared with the core, substantial material and engineering aspects. Also, the poor intellectual property regime in China not only leaves designers' work almost unprotected, but also prevents designers from establishing credibility and status that may enable them to confront other occupational groups with regard to design matters (particularly those art-related ones). Such a lack of status and credibility motivated designers to explore other approaches which might grant status or credibility. Although taking an engineering rhetorical stance did not help enhance designers' status, it did make designers' arguments more credible and less likely to be challenged, because engineering was virtually unbeatable in the company.

4.3 Implications for Design Work

How do designer identity and its evolution influence design work? Our preliminary data analysis has revealed three important ways. First, the evolution of designer identity was accompanied by a shift in the orientation of creative thinking. For newly minted designers, creativity was to a large extent attributed to an innate "gift for art" and the trainings that were necessary for effective expression of such a gift. They believed that the essence of design was creativity, and that the purpose of design was to create aesthetically pleasing and unique looks that convey particular meanings to the customers. As a result, these designers typically spent most of their time on browsing (various sources) for inspiration, sketching, and communicating with fellow designers. The time devoted to these practices decreased as designers matured, because designers became more and more preoccupied with engineering and business concerns which had been either absent or much less significant in their training at school. Designers became more and more involved in discussing, negotiating, and even fighting with engineers and product managers about issues such as structural feasibility, risk of failing a test, and cost of materials. In designers' own words, the creative element of work had been "squeezed out" or "corroded" by the time, quality, and cost pressures. Among more mature designers, the concept of creativity was not even primarily attached to the artistic dimension of design work, but was subsumed within the engineering and business frames. To them, creativity was more about figuring out a design solution that could simultaneously meet the cost and testing standards.

Second, the diminishing artistic component of designer identity led to symbolic use of certain design practices. Above we have shown the decline of the artistic component of the designer identity as designers become more and more experienced; and we explained that such a decline was primarily driven by organizational and institutional factors that prioritized pragmatic values (embraced by the engineering and business rhetorics) over idealistic ones (embraced by the art rhetoric). Despite such environmental pressures toward a depreciation of artistic values, designers nevertheless felt a need to champion the artistic aspect of design work, because it was the only thing that differentiated designers from other occupational groups. To survive in the organization, however, designers could not follow artistic pursuits in any substantial way; they had to primarily conform to environmental demands (for engineering and business values). Consequently, work practices that focus more on the creation of artistic values, such as beauty and novelty, were performed non-substantially; they became symbolic.

One such work practice observed in our study was brainstorming – a design work practice that is well known for its utility in generating *different and unique* ideas; in this sense, it fits the art rhetoric better than the engineering and business rhetorics (see Table 1). We noticed that designers were all very enthusiastic about brainstorming – they talked about it proudly and participated in it with great passion. We also noticed that the brainstorming sessions were indeed productive – many "special" and "unexpected" (in designers' own words) ideas were raised every time. However, brainstorming contributed little to the subsequent design proposals – few of the ideas generated in brainstorming sessions were actually adopted and developed into design proposals. In one case where designers brainstormed about accessories for a newly-developed smart phone, a list of 21 unique ideas were generated within 45 min, including, for instance, a protection case that could also stand the phone in either a vertical or horizontal position, an adjustable holder that makes it easier for people to use the phone when they lie on the bed, and a dock that connects the phone to a TV. However, on the next day, the design director brought in a bag of smart phone accessories that had been selling well on the market and the designers collectively selected three (one simple leather case, one simple plastic case, and one charger dock) to guide their design. Ideas from the brainstorming session were completely abandoned. Reflecting on the use of brainstorming, one designer said, "We all love brainstorming because it is where you can truly open up your mind and stretch your imagination . . . [But] most of those ideas are not feasible, at least in here (the company), because people will [either] tell you they don't know how to make it . . . [or] tell you they think it would be a market failure."

Finally, designers with a less mature identity were buffered from direct interaction with clients (the product management teams within the company that initiated design projects). What was striking to us during the early stage of our field study was the fact that most designers were not present in the initial design specification meetings (where clients explained what they wanted) and the final design proposal meetings (where design proposals were evaluated and selected). Design requirements and the evaluation results were passed to designers primarily by the design director who attended those meetings, and occasionally by the one or two veteran designers who were brought to those meetings by the design director. We were initially puzzled by such a buffering operation because we thought it would harm design work by preventing designers from a precise understanding of client requirements and a faithful explanation of and defense for their designs. But we later realized that this was not the case. Such a buffering operation proved to be conducive to design because it effectively eliminated the misunderstandings and conflicts that frequently arose in the interactions between designers and product managers, due to their incompatible ideologies of design. Product managers, mostly coming from a background of marketing or engineering management, subscribed more to the engineering and business rhetorics than to the art rhetoric. Less mature designers, on the contrary, primarily identified with an artistic ideology of design and were not quite used to an engineering or business way of talking about design. Designers and product managers might therefore misinterpret each other when a particular object or expression embedded different meanings specific to different rhetorics of design – for example, "strong" as a visual perception of masculinity (rhetoric of art) versus a sense of structural robustness (rhetoric of engineering). The buffering operation worked because the "bufferers" or mediators – that is, the design director and veteran designers – had a firm grasp of the alternative rhetorics of design, allowing them to appropriately interpret between different logics of design.

5   Conclusion

In this preliminary research, we have developed a unique approach for studying designer identity and its evolution. Through a lens of occupational rhetorics and using ethnographic methods, we have demonstrated that designers adopt distinct meaning systems to define the nature of design work and the design profession; and the use of these alternative meaning systems may exhibit a certain pattern as designers become more and more experienced in work. However, the specific findings reported here should be interpreted with great caution. First, this work is very preliminary and the insights presented are the result of an, as yet, superficial analysis of the data. Although we have confidence that the insights that we write about in this chapter reflect what we observed and what we were told by designers, more data collection and much more analysis is necessary to make sense of all that we learned and to derive deeper insights. Second, our insights are based on a close examination of a particular group of designers in a particular cultural context. Since the work of designers vary across cultures (Hinds and Lyon 2011), it is yet to be discovered how designer identity develop and change in other organizational, industrial, and national contexts.

References

Ashforth BE, Kreiner GE (1999) "How can you do it?": dirty work and the challenge of constructing a positive identity. Acad Manag Rev 24(3):413–434

Ashforth BE, Harrison SH, Corley KG (2008) Identification in organizations: an examination of four fundamental questions. J Manage 34(3):325–374

Carvalho L, Dong A, Maton K (2009) Legitimating design: a sociology of knowledge account of the field. Des Stud 30(5):483–502

Charmaz K (2006) Constructing grounded theory. Sage, Thousand Oaks

Fine GA (1996) Justifying work: occupational rhetorics as resources in restaurant kitchens. Adm Sci Q 41(1):90–115

Hinds P, Lyon J (2011) Innovation and culture: exploring the work of designers across the globe. In: Meinel C, Leifer L, Plattner H (eds) Design thinking. Springer, Berlin/Heidelberg, pp 101–110

Ibarra H, Barbulescu R (2010) Identity as narrative: prevalence, effectiveness, and consequences of narrative identity work in macro work role transitions. Acad Manag Rev 35(1):135–154

Snow DA, Anderson L (1987) Identity work among the homeless – the verbal construction and avowal of personal identities. Am J Sociol 92(6):1336–1371

Strauss AL, Corbin JM (1998) Basics of qualitative research: techniques and procedures for developing grounded theory, 2nd edn. Sage, Thousand Oaks

Stryker S, Burke PJ (2000) The past, present, and future of an identity theory. Soc Psychol Q 63 (4):284–297

Swann WB (1987) Identity negotiation: where two roads meet. J Pers Soc Psychol 53 (6):1038–1051

van Maanen J, Barley SR (1984) Occupational communities: culture and control in organizations. Res Organ Behav 6:287–365

Wang D, Ilhan AO (2009) Holding creativity together: a sociological theory of the design professions. Des Issues 25(1):5–21

AnalyzeD: A Virtual Design Observatory, Project Launch Year

Martin Steinert (Co-I), Hai Nugyen, Rebecca Currano, and Larry Leifer

Abstract This chapter describes the launch year activities of the analyzed project where we aim to quantify engineering design behavior to such an extent that use statistical algorithm to discover, describe and model fundamental design thinking behavior paradigm. It is a joint research endeavor with the EPIC chair of Prof. Hasso Plattner at the Hasso Plattner Institute (HPI), University of Postdam. As main result from the Stanford side, we were able to generate several proofs of concepts on gathering and analyzing design process data from various sources and in various data quality. Especially noteworthy is the analysis of CAD data in a novel and comprehensive way. Collaborating with a leading CAD software supplier, we are able to firstly extract every single engineer-system interaction and secondly, using genetic algorithms, we are able to statically identify patterns without an a priori model assumption.

1 Analyzed: The Project Aim

As described in (Skogstad et al. 2009) the main problem of modern design research is the missing of a quantitative general database that allows the comparison of various design processes independent of product or complexity.

With AnalyzeD we aim to create and disseminate a design project analyzer that will enable researchers beyond the HPDTRP community to conduct Design Thinking research. The application will be set up as software as a service (SaaS) solution. As a result we remain in partial control of the ongoing research activities. That will allow HPDTRP to benefit directly, by having data access and indirectly, by being cited. The initial setup of the service will encompass the functionalities as

M. Steinert (Co-I) • H. Nugyen • R. Currano • L. Leifer (✉)
Center for Design Research (CDR), Stanford University, School of Engineering,
424 Panama Mall, Stanford, CA 94304-2232, USA
e-mail: leifer@cdr.stanford.edu

H. Plattner et al. (eds.), *Design Thinking Research*, Understanding Innovation,
DOI 10.1007/978-3-642-31991-4_11, © Springer-Verlag Berlin Heidelberg 2012

developed for the d.store application within the last 2 years but based on generic architecture that offers simplified access and is able to handle large data sets. Beyond existing d.store functionality, analyzeD will allow us to tap into CAD log file data of various engineering projects. Equipped with this empirical data, we aim to quantitatively model and statistically test Design Thinking paradigms. A strong candidate for testing is the consecutive rapid iteration paradigm. Can recurring patterns and wave-like movements in the captured design activities indicate a well-explored solution space and enhanced output quality? To our current knowledge, testing core Design Thinking assumptions with large, real life data samples would be a first in Design Thinking research.

Whereas the HPI team of the EPIC chair focuses on enhancing the d.store analyzer interface, the CDR Stanford team has been concentrating on design paradigms and ways of quantitatively measuring design behavior or activities.

2 Proof-of-Concept: Capturing Designer Activity Data

In order to understand the viability and possibility of capturing temporal, i.e. longitudinal activity data from designers and having them ranked by means of a quality criterion, we have decided to follow our own design thinking methodology and ran a couple of iterations in class. Since our ultimate goal is to interface with CAD data, we opted to follow and analyze a course at the Product Realization Lab (PRL) at Stanford. The PRL is a teaching facility affiliated with the Mechanical Engineering Department. This lab supports design, creation and innovation through traditional machining, woodworking, foundry, plastics molding, welding, finishing, and metrology tools. State-of-the-art computer-aided drawing, manufacturing, and prototyping systems are also available.[1]

We were allowed to accompany "ME 318 – Computer-Aided Product Creation" and prototype design activity data gathering within the course.

Computer-Aided Product Creation – Prototype design and fabrication emphasizes the use of computer-supported tools in the design process. Students choose, design, and build individual projects using computer numerical control (CNC) software and CNC milling machines.

This course is especially interesting as it incorporates all stages: ideation, design, CAD, and production and as it produces 6 final prototypes during the course per participant. In total, we iterated and tested 6 versions of our questionnaire, attempting to use a web interface but quickly switching to a paper-based process. It became apparent that asking designers to provided information about their activity, even when confronted with a very simple input situation, does not produce reliable

---

[1] For more information, please visit: http://www.stanford.edu/group/prl/prl_site.

INSTRUCTIONS: *Please fill out this design activity sheet every day only what you have done,*
if no activities took place, please just date and sign.
(We are trying to quantify and understand the design process)

ACTIVITIES	TYPE OF CHANGES	ITERATIONS/TEST (number)	TIME (h:min)

Ideation
(non CAD, non final material)   modular architectural

number of iterations	time spent
	:
	:

CAD design   modular architectural

number of iterations	time spent
	:
	:

Prototyping
(final material)   modular architectural

number of iterations	time spent
	:
	:

Testing/ Simulating   modular architectural

number of tests/simulations		time spend on tests/sims	
successful ones	unsuccessful ones	successful ones	unsuccessful ones
		:	:
		:	:

Please indicate your design activity in terms of **radical vs. incremental**: _____ % to _____ %

one sentence on what you did today and why

free text: _____

Fig. 1 Fifth iteration of a survey setup to quantify prototyping behavior during the design process

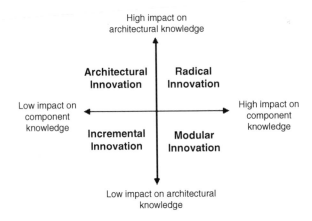

Fig. 2 A conceptual framework to identify and classify prototyping activities (Henderson and Clark 1990)

information. The last, and most simple questionnaire comprised the following (Fig. 1):

Additionally we provided the Henderson-Clark framework to explain our methodological quantification of the design process on the one hand and the resulting innovations on the other (Henderson and Clark 1990) (Fig. 2).

The data feedback, the observations and the post project discussions with the course instructors, teaching assistants and students clearly show that any data capturing process that relies on the designer to provide input, even if this is just a click at a key moment (say, every time he/she has created a new prototype) does not produce reliable data. The two main problems are, firstly not every iteration activity/act is being recorded and secondly, even using closed and quantitative survey questions combined with an established definition framework does not eliminate a subjective understanding of what comprises an iteration in terms of starting and end point and in terms of activity kind: e.g. modular versus architectural and

Hence we conclude that the data acquisition process must

1. Be completely automated and
2. Be non-invasive and invisible for the participating engineers.

3   Proof-of-Concept: Capturing Fuzzy Front-End Creativity Design Activities

In order to understand how much of the creative process we can reasonably target for quantification, we have explored the fuzzy front-end of the design process as extreme prototype, reflective practices.[2]

The "fuzzy front-end of innovation" offers the context for laying out the foundation for innovative design through new concept creation (Kim and Wilemon 2002), yet this front-end is poorly understood and therefore offers significant opportunity for improving the innovation process (Koen et al. 2001; Reinertsen 1999; Brown 2008; Plattner et al. 2009) and shows Design Thinking to be a successful approach in encouraging new concept generation early in the design process. In combination with project-based learning, some general Design Thinking practices have also found their way into the classrooms or lofts (Dym et al. 2006).

However, the practical means to promote increased creative ideation remain elusive and poorly understood both in practice and in education, and research has offered little in the way of impacting this. The move toward project-based learning and problem-based learning in engineering education presents an appropriate and timely opportunity to understand, transfer and incorporate reflective practice from design practice into engineering education.

The starting point for our research is a situation known to all engineers and designers: when getting stuck on a certain problem it is sometimes best to lay it aside and engage, maybe purposely, in a completely unrelated activity, such as

---

[2] This paragraph is based on the original work of this chapters authors: Currano et al. (2011), Currano and Steinert (2011).

taking a walk or showering. Quite often, these activities, which we will call reflective practices, suddenly generate new insights or open up new solution pathways. This phenomenon of reflective practice is at the center of our research. Our general guiding questions were:

1. What role does reflective practice play in Design Thinking?
2. How can designers learn to use reflective practices purposefully to maximize the potential for creative ideation?
3. How can reflective practices be integrated into engineering education?

The analysis of the interview data has unearthed some key reflective practices:

3.1 *"Mindless" Activities*

We were intrigued to find that two graduate participants specifically mentioned mindless activity as a basis for reflective practice. The third echoed this notion with the statement that "you need to not be doing anything too mentally taxing" and described napping as her primary reflective practice, saying "I'm actually a huge napper – all of my good ideas come when I'm napping" and "I'm one of those people who, when I'm kind of in that half-asleep but not quite asleep state, gets all of my good thinking done." One undergrad participant said, "I don't have to focus on anything except what I have to think about and if I wanna think about nothing that's possible too. And like, sometimes ideas just sort of creep into my mind when I'm in the rhythm of running."

3.2 *Physical Activities*

All three grad students and one undergrad cited exercise (going to the gym, running, biking) as a reflective practice, and the remaining undergraduates all included the physical activity of walking among their reflective practices.

3.3 *Conversation*

Talking to friends, in person or over the phone, was another commonly mentioned reflective practice. One undergrad described talking with her friend while driving and bouncing random ideas off each other. Another mentioned frequently talking to his housemates and getting ideas from their conversations. Still another described two different kinds of conversation as reflective practices – "group thinking conversations" with his project teammates and dinner conversations with others.

3.4   Remembering

One graduate student and one undergraduate especially emphasized remembering ideas. The first noted that his doodles (one of his primary reflective practices) all hold for him strong connections to the memory of whatever he was listening to at the time. The second said, "I'll have these great ideas and I'll never write them down, so it's just forgotten." Still another grad student referenced memory less directly, saying that while biking "stuff just kind of like filters through and you kind of rethink through conversations you've had and different things'll pop out, like 'what did they mean when they said that' or 'maybe you could put those other two ideas together and something cool would come from it.'"

3.5   Sketching

On graduate student indicated doodling as a reflective practice and said that he doodles a lot. Other students generally noted a value in sketching while talking about their reflective practices, but either said that they didn't do it enough, or that they struggle with sketching and can't sketch fluidly enough.

3.6   Validating the Reflective Practice Framework

When overlaid on the same graph we can see a broad distribution of reflective practices and roughly equal representation in all four quadrants (Fig. 3). This distribution, resulting from participants' mapping of pre-described practices in the framework, indicates the appropriateness of the dimensions in characterizing reflective practice. This is further supported by the fact that participants, while sometimes hesitating or confusion in initially identifying or naming their reflective practices, mapped them onto the framework with relative ease.

We did not find any inconsistencies or surprises in how participants mapped reflective practices on the y-axis (in-action/out-of-action) but we did note that different participants mapped walking and brainstorming on opposite sides of the remembering/gathering axis. This does not necessarily denote a flaw in our framework, as different individuals may approach the same general practice with a different focus, one focusing more on remembering and one more on gathering.

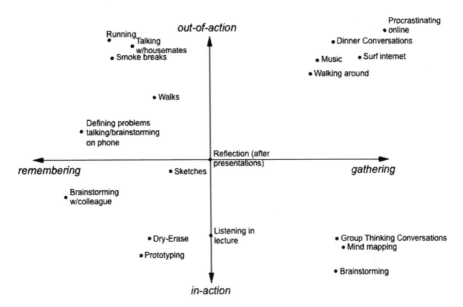

Fig. 3 Conceptual framework for reflection, populated by interview participants' reflective practices (Currano et al. 2011)

It might, however, point to confusion or discrepancy in their understanding of the meaning of the dimensions represented in the framework. We were surprised to find "mind mapping" mapped so far to the right (indicating a strong gathering component and a weak remembering component). But we do recognize both components at play in this reflective practice, as the designer maps what is already in his mind (memory component) but also may see things differently once they are on the paper and map new things that come from seeing his thoughts on paper (gathering).

Also notable, is the higher number of practices mapped to the peripheries than to the middle section of the framework. A possible explanation might be that the design students purposely, or simply by experience, engage in practices that combine the two proposed dimensions in order to profit from reflective practices.

Based on these finding we have developed and pilot tested a web-based survey to run a factor analysis onto the general dimensions of reflective design practices. We are currently attempting to gather large-scale data sets using Mechanical Turk, the online crowdsourcing marketplace from Amazon.

This stream of research has been successfully submitted and presented at two peer review conferences, the eighteenth International Conference on Engineering Design, ICED11, 15.–18.08.2011, Copenhagen, Denmark and MDW VIII, the eighth Mudd Design Workshop, 26.–28.05.2011, Claremont, CA, USA. Furthermore, an elaborated version of the latter has been accepted at the Journal of Engineering Education (IJEE), to be published in 2012 in the Special Issue on Design Education: Innovation and Entrepreneurship.

4 Proof-of-Concept: Identifying Coordination Patterns in Multidisciplinary Teams

A second exploration aimed at understanding how participants in product design and development projects need to interact to coordinate the impacts and dependencies of their work on the product.[3] We analyzed data from NASA Mission Design Center projects that allowed us to compare the stability of established communication patterns across multiple design projects. Projects that address similar design problems show similar communication patterns. Based on this relationship, it is possible for program managers to optimize and support communication connections and to speed up organizational learning.

When a team takes on several projects, the observed connections among the participants might be similar or distinct when we compare the projects. A connection that was made in one project may or may not be made in another project. The experience acquired in a previous project may tell whether a specific connection is or is not relevant, what issues were identified and, perhaps depending on the nature of the product, if the set of connections are or are not different.

In order to test this, we formulate the following hypothesis:

Hypothesis

> The communication channels that are relevant in one project will be the same in a another project if these projects are similar

Measuring project similarity and similarity of communication importance in 13 NASA projects allowed us to compare the data sets. Additionally, we compared the result of random generated projects in order to check whether any statistical correlations were a random effect or based on an underlying correlation. Figure 4 depicts the result.

From these results we can observe that there is a direct correlation between functional similarity and the pattern of important communication channels. It is also observable that this correlation is not simply due to the constitution of the teams. Considering that participants in these projects are free to engage with any other participants, in terms of team interactions these results illustrate that similar projects have similar communication channels. When a team starts another project, its participants will mostly interact with the same areas as before, given that the new project is similar to the previous one.

This research, accepted and presented at the eighteenth International Conference on Engineering Design, ICED11, 15.–18.08.2011, Copenhagen, Denmark, shows that the quantification and analysis of abstract team interaction data allows uncovering underlying design paradigms.

---

[3] This paragraph is based on the original work of this chapters authors: Castro et al. (2011).

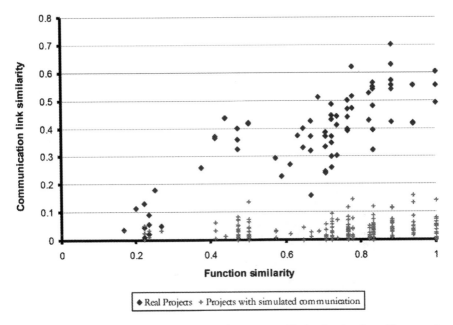

Fig. 4 Function similarity and communication importance with simulated projects (Castro et al. 2011)

5 Proof-of-Concept: CAD Log Data Analysis

In order to capture CAD activity data we needed to choose the appropriate application. After a thorough comparison between the market leaders, we decided to focus onto the leading high-end application, *anonymous* and the leading single engineer application, *anonymous*.

Initial contact was made, followed by numerous workshops in Stanford and Paris. In turns out the future applications, currently under development, are including ever more aspects of the design process such as collaboration and sharing, communication, pattern saving and reusing etc. Hence our aim to choose CAD data as primary input source seems to be very promising as it will entail ever more of the design process, including sketching and fuzzy front end activities (Fig. 5).

We were able to measure every single activity that a designer goes through within a CAD software during a design process (Fig. 6).

These time-stamped activities allow comparisons between designers and projects (Fig. 7).

AnalyzD data parsed from CAD log files thus allows us to tackle, for example, the following questions:

How does the designers' time distribution affect the project outcomes?
How do designers spend their time overtime?

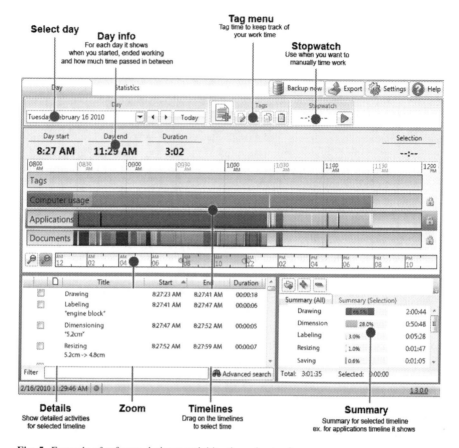

Fig. 5 Example of software designer activities throughout a day

Fig. 6 Example of CSV time-stamped activity data

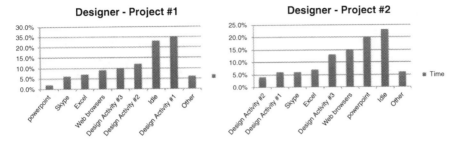

Fig. 7 Inter-designer/project comparison of design activities

Currently we are analyzing actual data provided by a CAD software company, see following paragraphs.

6 Proof-of-Concept: Software Design Log Data Analysis

Since the IP and legal issues have slowed our formal collaboration with the CAD application somewhat, we have chosen to also include software design. We opted for an automated capturing process and chose the software development environment Eclipse as test bed. The aim was to capture the activity data over time. As data source, we are using data of the SV company Appfluence. Appfluence LLC is a productivity software company based in Palo Alto, CA. It was founded in 2010 by four technology experts with ties to both the academia and industry elite of Silicon Valley. Appfluence has been invited to showcase their products in various events at Stanford University, such as the Stanford Product Showcase and the BASES product development accelerator Forge. Earlier this year, they hosted a booth at both MacWorld 2011 and mobile conference AppNation in San Francisco (Fig. 8).

The algorithm patterns the search behavior of 30,000 user sessions and allows identifying relationships and families. We are also in discussion with a large Silicon Valley Internet company to obtain clearance for using their programmers 'activity data. The proof of concept for anonymous was successful but may not be disclosed.

7 Proof-of-Concept: Statistical Analysis Logic and Algorithms

Our first year probing into potential algorithms and statistical analyses indicated two problems:

1. A priori knowledge of underlying principles and relationships and
2. Computational limits due to data size.

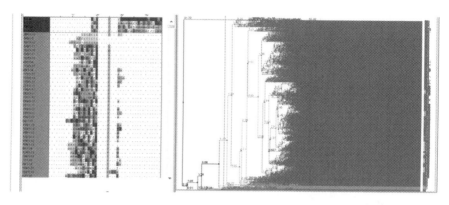

Fig. 8 Screenshot of Appfluence LLC. data analyzed by an bioinformatics inspired algorithm that was originally developed to investigate DNA sequence data. We investigated the sequences of events their customers go through

Standard statistical methods, ranging from simple correlation testing to causality based multivariate analyses, such as regressions analyses or structure equation model, for example, based on AMOS or Partial Least Squares (PLS), require the analyst to predict and hypothesize an underlying relationship that can then be tested. Based on that, the researcher can iteratively improve the concept and model fit. The drawback to this procedure is that the data must be preprocessed and limited to the hypothesized conceptual components.

However, if we process the entire data generate by the source systems, depending on the analyses used, we soon hit a limit or raw computational power that goes beyond simple desktop systems.

These have been long standing issues in the field of bioinformatics. This field has developed a number of algorithms and approaches to deal with data sizes in vast excess of brute force computation (Edgar 2004; Stamatakis 2006; Li and Godzik 2006). These approaches are often directed at problems where a priori knowledge of the system is limited. Since both biological and industrial entities are in constant competition for survival and resources, many of the concepts behind natural selection are applicable.

We propose to encode design and procedural parameters in a way that enables their analysis using existing tools that were originally developed to investigate DNA sequence data. These methods will allow clustering of design approaches by similarity, alignment of similar functions, and correlation of specific components of design to success or failure of designs as a whole. This will allow for a better understanding of the tolerance to change for certain aspects of a procedure, and charting the evolution of a design procedure, to see how ideas originate, and are adopted and modified over their lifetime. Figure 9 shows an example of an alignment algorithm on protein data. Each line represents a single protein sequence related to the others. The characters in the sequence are colored by their chemical properties. Proteins consist of a small set of smaller components assembled in a linear chain. As proteins evolve between organisms, insertions and deletions occur

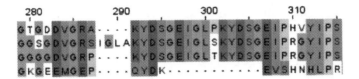

Fig. 9 An alignment of protein sequence. Components with similar properties share the same color

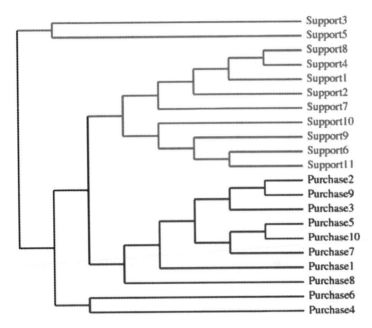

Fig. 10 A tree built by encoding user sessions as protein code. *Red* colored branches involve the seeking user support, while *blue* involves a purchase link. The algorithm partitions the tree according to similarity of user sessions

in the sequence, which make it difficult to directly compare regions. The alignment algorithm inserts gaps into sequences in order to define regions, which are directly comparable.

As a test to the viability of this approach, we encoded over 30,000 user sessions on the website of the company Appfluence, as though they were elements in a protein sequence produced from DNA. We took 21 user sessions as an initial test set. The tree in Fig. 9 shows how related or similar the sessions are to each other. We have colored sessions in which the user accessed the "User Support" page in red and those where the "Purchase" page was accessed in blue. The tree-building algorithm accurately partitions the two types, identifying different types of user experiences with the page. The separation of the branch with "Support3 and Support5" indicates that the users in those sessions had a very different browsing session than the other support users (Fig. 10).

8 Current State and Future Research

Having tested all the different critical sub functions of the research we are currently collecting more CAD data. Additionally we are planning to produce CAD data in a semi-controlled lab environment in which we can simulate certain design thinking paradigms and behaviors. Another two peer reviewed conference paper from this project year has been submitted and is under review.

References

Brown T (2008) Design thinking. Harv Bus Rev 86(6):84–92

Castro J, Steinert M, Seering W (2011) On the stability of coordination patterns in multidisciplinary design projects. In: Proceedings of the 18th international conference on engineering design (ICED11), Copenhagen, Denmark, vol 3, pp 245–252

Currano RM, Steinert M (2011), A framework for reflective practice in innovative design. In: MDW VIII, Mudd design workshop, Claremont

Currano RM, Steinert M, Leifer LJ (2011), Characterizing reflective practice in design – what about those ideas you get in the shower? In: Proceedings of the 18th international conference on engineering sesign (ICED11), Copenhagen, Denmark, vol 7, pp 374–383

Dym CL, Agogino AM, Eris O, Frey DD, Leifer LJ (2006) Engineering design thinking, teaching, and learning. IEEE Eng Manag Rev 34(1):65–92

Edgar RC (2004) MUSCLE: multiple sequence alignment with high accuracy and high throughput. Nucleic Acids Res 32(5):1792–1797

Henderson RM, Clark KB (1990) Architectural innovation: the reconfiguration of existing product technologies and the failure of established firms. Adm Sci Q 35:9–30

Kim J, Wilemon D (2002) Strategic issues in managing innovation's fuzzy front-end. Eur J Innov Manage 5(1):27–39

Koen P et al (2001) Providing clarity and a common language to the 'fuzzy front end'. Res Technol Manage 44(2):46–55

Li W, Godzik A (2006) Cd-hit: a fast program for clustering and comparing large sets of protein or nucleotide sequences. Bioinformatics 22(13):1658–1659

Plattner H, Meinel C, Weinberg U (2009) Design-thinking. Finanzbuch Verlag GmbH, München

Reinertsen DG (1999) Taking the fuzziness out of the fuzzy front end. Res Technol Manage 42 (6):25–31

Skogstad P, Steinert M, Gumerlock K, Leifer L (2009) We need a universal design project outcome performance measurement metric: a discussion based on empirical research. In: Proceedings of the 17th international conference on engineering design (ICED'09), Stanford, USA, vol 6, pp 473–484

Stamatakis A (2006) RAxML-VI-HPC: maximum likelihood-based phylogenetic analyses with thousands of taxa and mixed models. Bioinformatics 22(21):2688

When Research Meets Practice: Tangible Business Process Modeling at Work

Alexander Luebbe and Mathias

Abstract We have created a modeling approach used by people in organizations to create and discuss business process models that represent their working procedures. This is an alternative to established approaches in which process modeling experts create business process models for the organization based on input from domain experts. We have changed this by empowering the domain experts to model their business processes themselves. This approach consists of a simple to use haptic toolset and the facilitation for its application.

In the first stage of research, we showed that our approach, called tangible business process modeling (t.BPM), can be used to co-create process models with novice modelers. In a subsequent laboratory experiment, we found that t.BPM is superior to interviews for process elicitation because people are more engaged with the modeling task and the result is better validated. Furthermore, people have more fun and develop a better understanding of the process.

In current research, we developed and assessed the idea of t.BPM for application in professional environments. We are seeking to change the state of business by showing the feasibility of t.BPM for real modeling projects. We investigated when and how to apply t.BPM correctly. In doing so, we were able to show that t.BPM is mature enough to compete with established workshop techniques.

A. Luebbe • Mathias (✉)
Business Process Technology, Hasso-Plattner-Institute, Potsdam, Germany

H. Plattner et al. (eds.), *Design Thinking Research*, Understanding Innovation,
DOI 10.1007/978-3-642-31991-4_12, © Springer-Verlag Berlin Heidelberg 2012

1 Introduction

Business processes consist of interrelated tasks that are carried out in coordination to reach a business goal. They implicitly exist in each organization. When making implicit processes visible, one can describe them as a model. Creating such a model for business processes is called business process modeling. It is the act of visualizing the knowledge about work procedures in an organization. Business process modeling is a common activity in business process management, which is a holistic approach to structure, measure and coordinate work in organizations. The idea is to investigate, change and monitor procedures that drive the daily business operations.

The process model is an artifact used in discussions about the business process to form a shared understanding among stakeholders of the process and discover opportunities for improvement of the organization. Many people have a stake in business process modeling, e.g. the managers who gain an overview about the current state of business operations and decide about future alternative ways of working. Furthermore, public accountants audit whether processes are performed as defined in the organizational handbook and as required by legislation. Finally, software engineers have become important stakeholders in business process modeling. When work shall be automated in software systems the business process models describe how systems and people interlink.

Since software is omnipresent in today's organizational environments, changes in the process often affect software systems that need to be changed as well. Supporting business processes with software systems yield great potential to save time, enhance reliability and deliver a standardized result. The communication between business and IT departments is therefore crucial to the success of the overall business. And today, this communication often relies on business process models.

1.1   Current State of the Art

At present, process modeling is a special skill. Typically, business analysts get trained in process modeling and sell this skill as a consulting business. They interview the stakeholders of the process and create a model that reflects their understanding of the organizational procedures. The model is created using a specialized software tool. Various expert tools have evolved for that purpose; see Fig. 1 as an example. In any case, the modeling tools remain in the hands of the consultant for efficient use. To gather feedback from domain experts the process is printed and passed back. However, domain experts often do not sufficiently understand the models or decide that their knowledge is not well represented. Additional effort is needed to explain the model and correct mistakes. Three to five iteration

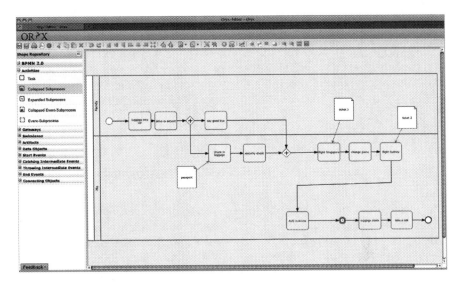

Fig. 1 A sample business process model in the BPMN (OMG2011) language. The model was created with the Oryx-Editor, a software modeling tool for business processes

cycles are common in reaching a consensus about the final model, in other words, to create a shared view.

1.2   When the State of the Art Is Not Enough: One Example

We illustrate the relevance of this topic by the example of one specific company. LISA corp. was a federal funding agency where legitimate environmental projects applied to receive funding. LISA assessed projects, paid funding and claimed refunds from European authorities.

In 2006, LISA decided to purchase new software to support their core processes in the area of funding and refunding. They hired an external consultant, who was a process modeling expert, to prepare a call for tenders. The consultant carried out interviews with clerks from LISA, the domain experts. As a result, he created a set of business process models describing how LISA corp. will work in the future provided with new software. The process models were used as part of the call for tenders, the configuration of the installed software and, finally, in the organizational handbook to document the new work procedures of LISA. The clerks were trained to work accordingly. The new process was established in 2008.

Unfortunately, there was a significant error in the process models. Payments might go to non-legitimate projects because prerequisites were not checked adequately, neither by the clerks nor by the software system. Nobody realized this was happening until another stakeholder investigated the models in detail: The auditor from the European funding agency finally discovered what was going on.

LISA corp. did not receive further refunds for already funded projects. In 2010, LISA corp. went out of business.

Besides mistakes that might have happened this way, there often is a miscommunication between two stakeholder groups, the domain experts and the process modeling expert. This is one example of why capturing knowledge in process models is three critical task and why the current state of practice is not enough.

1.3   The Idea of Tangible Business Process Modeling

This situation was the starting point for our research three years ago. We wanted to address the flaw of limited user empowerment that could lead to misunderstandings, as in the case of LISA. We started by prototyping ideas for model building together with the domain experts (Edelman et al. 2009; Grosskopf et al. 2009; Plattner et al. 2010). We developed an approach that we call tangible business process modeling (t.BPM). It consists of a toolkit and a method for its application.

The toolkit is a set of tangible shapes reflecting four basic icons of the BPMN (Business Process Model and Notation) (OMG 2011). This is an international standard for business process modeling supported by more than 70 tools.[1] The concepts of the notation are embodied in thick acrylic tiles that are laid out on a table and can be inscribed with whiteboard markers. The tiles represent the concepts of activities (boxes with rounded edges), events (circles), routing nodes (diamond shapes) and the information used in the process (document shapes). The actual flow of the process steps is marked on the table with whiteboard markers as well. The same holds for role information. It is drawn on the table as swim lanes and activities placed in a swim lane are to be performed by the assigned role. Figure 2 depicts a process model embodied in t.BPM. It is the same process model as the one in Fig. 1. It shows that t.BPM is a low tech tool for process modeling that has the same expressiveness as high fidelity software tools.

Software tools yield advances such as seamless replication and simulation of models, but they also manifest a barrier between those that create models and those that consume the result. The idea of t.BPM is to overcome these barriers and enable more people to engage in the creation of process models. We envision t.BPM to be used in workshops by domain experts to create and discuss process models.

The role of the process modeling expert changes from a model creator to a facilitator. He explains the concepts of process modeling to workshop participants and guides them through the steps of model creation. Instead of abstracting and framing the information for the domain experts, he helps them to do it themselves. His modeling expertise is still needed for difficult modeling situations. He must ensure that the modeling rules are not violated and that the model is relevant to the

---

[1] http://www.omg.org/bpmn/BPMN_Supporters.htm

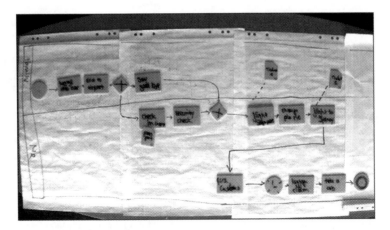

Fig. 2 The same business process as in Fig. 1. This time the process was modeled with the tangible business process modeling (t.BPM) toolkit

context that it is to be used in. Yet, more importantly, he is a moderator to the group at the workshop.

2   Previous Research and the Big Picture

We started with the observation that domain experts are not directly involved in process modeling. That spawned a series of prototypes to test ideas that familiarize new audiences with process modeling. The prototyping phase was guided by scientific knowledge from related disciplines, such as cognitive science and design thinking research. We learned about mental effort of learners (Sweller 1988) and affordances of media in steering design discussions (Edelman 2009). We explained this initial part of the journey in a chapter found in the first volume of this book series (Plattner et al. 2010, 2011). Our result was the idea of tangible business process modeling as a promising solution. At that time, there was little understanding about why t.BPM actually worked and if it was a realistic solution, practically speaking.

In the next stage we conducted a series of case studies (Luebbe and Weske 2010) that helped us to build knowledge about the idea and further refine it towards a solution. At that stage we iterated the tangible toolkit and built first insights about the method for t.BPM application. Afterwards, we assessed hypotheses about the t.BPM toolkit in a laboratory experiment. In that experiment, we compared t.BPM with structured interviews. So, people either spoke about their process or modeled it using our tool. We found by video analysis and questionnaires that tangible modeling yields advantages. In particular, people review and correct the models more often leading to better validated process models. They were also observed to be more engaged with the task and spent more time talking and thinking about their processes. The test participants also reported that they enjoyed the t.BPM modeling

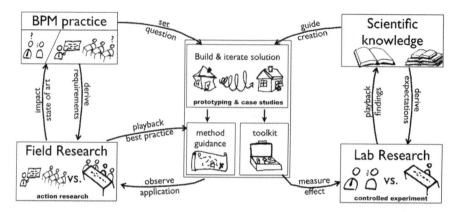

Fig. 3 The big picture of the t.BPM research efforts. Structuring the solution and investigating it in a lab experiment were chapter themes in the first volumes of this book series. This book chapter focuses on field research

more than interviews and that they gained new insights about the process. The experiment was analyzed from various angles and the results were published in smaller pieces (Luebbe and Weske 2011a, b; Luebbe 2011). A summary of the experiment was contributed as a chapter to the previous volume of this book series (Plattner et al. 2011).

The experiment was an important step towards understanding the effects of tangible modeling on the participants. Still, it did not yet provide method guidance for people who want to use the tool for group workshops. Our goal hence is to bring the advantages of t.BPM, found in the lab experiment, into practical use. We want to improve the way processes are communicated between the stakeholders in today's business environments. Thus, we undertake field research as the next step. This means going out to real process modeling projects and applying t.BPM there in order to learn about all the context dependencies that we did not see in the laboratory experiments with individuals only. In particular, the setups required to run such a workshop in-line with the facilities given are of primary interest. Figure 3 illustrates the different aspects of this research in a overview. We have built a solution and investigated it in the laboratory. Now, we conduct field research to further evolve it and validate the applicability in real settings. We want to impact the state of the art in business process modeling. In order to do that, the following research questions are formulated:

1. How are tangible process modeling workshops set up and facilitated?
2. When are tangible modeling workshops favored over other process elicitation techniques?
3. How does the t.BPM result differ from other process modeling workshops?
4. How does the performance of t.BPM workshops compare to other process modeling workshops?

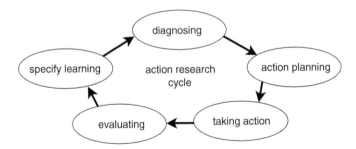

Fig. 4 Action research cycle (Adopted from Baskerville (1999) to guide our research collaboration with practitioners)

3 The Action Research Method

We decided on field research as our approach, in other words, research in professional environments with real stakeholders and real problems. We opted to collaborate with BPM consultants on their job. They should try out t.BPM while we act as observers, learn from them, and propose alignments according to scientific theory. This enables us to observe extensively and remain open for new insights from experienced experts. We choose action research (Lewin 1946) as the scientific method to guide us in the collaboration with practitioners. Action research is a class of research methods in which researchers collaborate with practitioners to act in or on a social system (Brydon-Miller et al. 2003). The goal is to solve practical problems and generate scientific knowledge. Action research assumes that complex social problems cannot be reduced to a meaningful controllable study. They need to be investigated within the context in which they appear. Therefore, the system is studied, changed, and the effect of change is studied again. This is an iterative process. Kurt Lewin described it as a spiral of "planning, action, and fact-finding about the results of the action" (Lewin 1946).

We have adopted a five-stage model proposed by e.g. (Baskerville 1999) as described in Fig. 4. The way this cycle is operationalized in studies varies (see also Susman and Evered 1978). We describe the operationalization of the phases in the following paragraph.

After introducing the context for each study, we start by diagnosing the problems in the given context. These problems become challenges to be tackled in the current action research case. In *action planning* we describe the desired future state and propose changes. Proposed changes may be based on existing scientific findings or – if not applicable – are created in collaboration with the practitioners. These changes are to be assessed for their expected effect. The first two research phases, *diagnosing* and *action planning*, are carried out in close collaboration with the practitioners. In *taking action* the practitioners act alone while the researchers observe and concentrate on data collection for the *evaluating* phase. In the evaluation, we assess whether the proposed changes realized the anticipated effect and whether the practitioners' problems were solved.

Finally, we step back from the specific case in *specify learning* to derive more general conclusions, practically and scientifically. This last phase determines knowledge and questions to carry over to other cases in subsequent iterations of the action research cycle. Each study performed in this chapter represents one iteration of the action research cycle.

Differences to consulting. Action research is sometimes mistaken as consulting which, in fact, it is not. There are established rules to follow in action research studies (Davison et al. 2004), such as the commitment to generate scientific knowledge. Likewise, the advice given by the researcher should be scientifically grounded. The commitment to science might even result in a smaller benefit for the practitioners for the sake of scientific insights (Baskerville 1997). These are fundamental differences to consulting which is purely client-oriented without commitments to science.

4   Action Research Studies

We conducted a series of action research studies with different partners from industry. In each case, we collaborated with a consultant who carried out a project and applied t.BPM. We followed the research cycles as explained in Fig. 4 to work out our research results. We briefly describe the studies in this section and show the results in the next section.

4.1   Hospital Doctors Modeling Clinical Pathways

In mid 2009, a BPM consultant approached us to use the t.BPM set in a professional workshop. We agreed and sent an observer to the workshop. It was held at a university hospital to capture and optimize clinical pathways with doctors. A clinical pathway is a treatment process in a hospital for a class of patients with a similar disease, e.g. a certain type of cancer. This was the first application of tangible business process modeling in group workshops. Within one week, the group modeled 20 BPMN processes. The workshop was considered a success by the BPM consultant. We published insights from this first t.BPM application at the Electronic Colloquium on Design Thinking Research (Luebbe and Weske 2010).

In mid 2010, the same consultant approached us again with the same request. The setting was very similar, only the participants and the clinical pathways were changed. Thus, the new workshop in 2010 would be an iteration of the workshop held in 2009. This was the starting point for the first action research cycle. We reviewed the workshop from 2009 and identified options for improvements. We discussed, among other things, a shorter introduction and modeling exercises for the start of the workshop. Therefore, we collected modeling exercises from practitioners and scientific literature such as 'withdrawing money from an ATM', inspired by Rittgen (2010). These scenarios enable participants to get familiar with process modeling without getting lost in domain specific discussions.

Fig. 5 Snapshots were taken at the 2010-workshop with the hospital doctors. The consultant (*left person on left picture*) taught process modeling by a hands-on modeling exercise to kick-start joint model creation

Again, a one-week workshop was conducted. Three participants from the hospital worked out the clinical pathways for liver transplantation together with the consultant. The doctors modeled 13 processes throughout the week. Figure 5 shows some impressions from the workshop. In total, we collected more than 600 pictures, 9 pages with observer notes, and interviews with the consultant and each participant. From the modeling sessions, we collected six snapshots of the project during the week documenting the process models at their current state.

We used the data in our learning phase to evaluate to what extent we could improve the situation over the previous workshop. The positive aspects were condensed in a first methodological guidance for t.BPM workshops. It explains how to setup and facilitate t.BPM workshops. We will present an overview of it later in this chapter as part of the research results.

4.2 Innovation Managers Modeling Idea Management Processes

In early 2011, we got in contact with an in-house BPM consultant working for a subsidiary of a large energy provider in Germany. The company is responsible for developing and testing new ideas for the energy market such as smart-home devices and e-mobility concepts. Within this company, the BPM department captures and improves organizational processes in workshops together with domain experts. A typical modeling workshop lasts for 3 h and is run by two people, a moderator and a software tool expert. While the moderator elicits information in a conversation with the participants, the tool expert translates the information into a process model on the fly. The computer screen is simultaneously projected to a wall. The participants can see the process model evolving and review it. We call this a software-supported process modeling workshop (Luebbe 2011). The process modeling team has used this technique for more than 600 process models altogether over the past years. They approached us because they had a need for a different elicitation technique in specific workshop situations. For us, the main difference to our previous t.BPM workshop was the modeling notation, which in this case was event-driven process chains (EPC) (Keller et al. 1992).

Fig. 6 Tangible EPC (Keller et al. 1992) modeling toolkit (*left picture*) uses plastic tiles and colored post-its to show different aspects on the process. It was the basis for co-creation of the model with the consultant (*right picture*)

Before we went into the workshop, we discussed current flaws and what benefits t.BPM could bring to the workshop. Together we also characterized which type of processes would benefit from using t.BPM and which ones would not. Additionally, we designed a tangible EPC modeling toolkit, as a variant of the t.BPM toolkit which is designed to be used for BPMN processes. Figure 6 shows the set that we designed to be used in EPC modeling workshops.

Within the company the department for idea management was identified as suitable. The department, formed one year earlier, had just grown from two to three employees. The processes were not modeled yet but they were looking for software to store, rate, and track their ideas. Two workshop sessions were scheduled on two subsequent days. The workshop objective was to create an overview of the existing processes in this department and to model the core process that would benefit from software support. Figure 6 shows a snapshot from the workshop situation.

To collect data, we had access to the processes from the existing organizational handbook. It serves as a reference point for typical processes captured by the same team of process modeling experts in the same company using software-supported workshops. For one modeling project, we were also able to trace the workshop effort for each model. This enables us to trace productivity of the modeling team in software-supported workshops. To collect data from the t.BPM workshop, we sent two observers to take photos and notes. We collected three pages of observer notes and 112 photos including four snapshots of intermediate the modeling result. We used the data to compare t.BPM workshops with software-supported workshops with respect to the modeling result and performance. In discussions with the consultants, we updated our methodology and characterization of workshops that benefit from using t.BPM. We present the results in the next section.

5 Research Results

The insights presented here are framed in relation to the research questions stated earlier. The data was collected in the two action research case studies described above. Additionally, the methodology guidance was discussed with further BPM consultants and researchers to form a stable opinion on the relevance of the information depicted.

5.1 How to Setup and Facilitate

The workshop guidance is encompassed in eight method cards covering the practical hints regarding the modeling project, modeling sessions and the process to be modeled. The cards were designed for experienced process modeling experts that are using t.BPM for the first time.

As examples, for the modeling project we propose establishing a media framework. This means to communicate the correct use of different media for different purposes. Thus, people should use t.BPM for co-creation of process models during the workshop and software to store the workshop results. Discussions should always be accompanied by drawings. The advice is based on the media models theory (Edelman 2011) that describes the steering effect that media has on the communication. It was validated by our empirical insights. Consultants and workshop participants told us they liked working out processes with the t.BPM tool and wanted to avoid 'fiddling with software in group workshops'.

To prepare sessions, we proposed inviting three to four stakeholders with different views on the process. They meet in a room with a large table and enough space to walk around the table. Conference rooms are well suited to their needs. The session is held while standing to bring a dynamic atmosphere into the workshop. It is important to communicate the scope of the modeling session as well as the rules to follow. Before the actual model is created a warm-up exercise, as explained before, might be used to practice process modeling without getting lost in content specific discussions. One method card is provided for sample warm-up exercises that may be used in sessions.

The process related advice in the method guide is concerned with modeling guidelines and facilitation of fast process mapping. Modeling guidelines are constraints beyond syntactic correctness of a model, such as the maximum number of elements that should be allowed per model. We propose to create compact models and create different refinement levels instead of producing detailed models. Furthermore, we observed in the workshops that a small number of modeling concepts already provides a notable amount of content. Introducing too many concepts confuses the participants about which ones to use when. Thus, we propose starting with a minimum amount and adding concepts iteratively. For example, we propose starting with start and end-point of the process. Afterwards, the main flow of work is worked out. Later the process model is enriched with documents and further concepts. To keep the shapes floating on the table, arrows between them are held off until the last possible point. Figure 7 depicts a t.BPM process in four phases of its creation.

Not all advice given in the method cards is specific to t.BPM workshops, but it is recommended when organizing good workshops in general, such as the prior

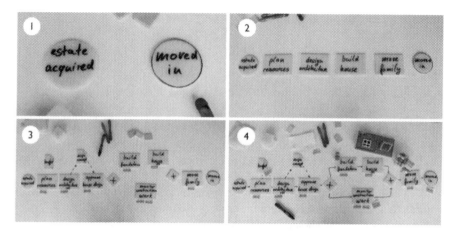

Fig. 7 A t.BPM process model in four phases of its development. First (**a**) the start- and end-point are decided upon and then (**b**) the main steps are mapped. Afterwards, (**c**) the process model is enriched with more modeling concepts and finally with (**d**) *arrows* and additional context information

scoping. We added information to the method cards when we encountered problems in our action research studies. Furthermore, BPM consultants and researchers rated the relevance of information in the method cards to form a useful set of advice. The method cards can be found as attachments to the dissertation on tangible business process modeling (Luebbe 2011).

5.2   When to Favor t.BPM

The t.BPM workshop method is an alternative to brown-paper workshops and software-supported modeling workshops. It might also be used as an alternative to interviews, but it was designed to facilitate groups that have a need to discuss their process models and shape the models together. This is typically the case for people who start process elicitation efforts for the first time. Additionally, radical redesign projects in which a new process is to be designed benefit greatly from this type of group workshops. If the process is not clear upfront because multiple stakeholders are required to create a big picture overview, t.BPM can help to create this overview. Similarly, the redesign of a process might require the integration of multiple stakeholders and the creation of a mutually agreed business blueprint. The objective of t.BPM workshops can be stakeholder integration, a shared common view, validation of or agreement to a plan. This is valuable especially when the organization is confronted with radical change.

Together with the consultants, we also identified counter indicators for t.BPM workshops. In particular, a well-established business process that does not require a radical redesign is not suited for a t.BPM workshop. In these cases, people know what they are doing and can talk very structured about their work. Often, the process is supported by IT-systems, which already automate parts of the process. This is often the case for core processes in well-established manufacturing industries such as the automotive industry.

Choosing one workshop method over another is a situational choice that also depends on the preferences of the consultant, the participants, and the company culture. We concentrated on the process to be modeled as the key criterion.

5.3 Comparison of Modeling Results

We collected tangible modeling results in all action research studies. In the second study, we also had data from other workshops to compare it with. We received access to the organizational handbook of the company that we worked with. It contained hundreds of models most of which have been created by the same consulting team using software-supported workshops. By characterizing these models and putting them in relation to the modeling results from the t.BPM workshop session, we can see differences of tangible modeling results to the other process models. We choose 22 processes from a similarly structured department to compare with our models. Figure 8 illustrates the results in a boxplot. A boxplot diagram (Field 2009) shows the distribution of data within different ranges. It visually partitions the data set into four equally large sets of data describing the 1st, 2nd, 3rd, and 4th 25 % of the dataset.

We characterize the models by quantifying the type of information embodied. The boxplot shows that 75 % of the evaluated processes contain 8–54 control nodes. Another 25 % of the models have 55–96 control flow nodes. This is core information such as activities and routing nodes in the model. We see that the tangible modeling result (the black square) is pretty much at the median indicating it is a truly average model in size. In other words, our tangible model was quite average in the number of steps modeled. That is an important finding because it suggests that tangible modeling does create properly sized models. Our model also had rich information about documents, IT systems and roles. While documents and IT systems are just slightly higher than average, role information in our tangible model was out of bounds. When investigating this, we found that the idea management department plays an integrative role. That means, they collaborate with many other departments within the organization, they frequently talk to the management, and they also collaborate with people outside the organization. That made this process model rich in role information. Additionally, the models in the organizational handbook often missed role information. Some models even had no responsibilities attached to the activities leading to zero role information in the process model.

We came to the conclusion that tangible process modeling does not create smaller or bigger models in terms of control flow information. It does, however,

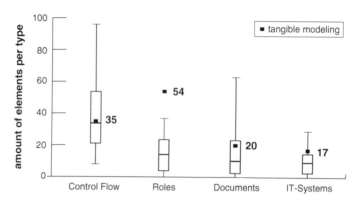

Fig. 8 The boxplot diagram depicts the distribution of process model characteristics in the existing organizational handbook. The *additional black squares* depict the characteristics of the tangible EPC model created during the workshop. It shows that our process model is quite a normal model despite rich role information

have enriched context information such as more roles, documents and IT-systems depicted in the models.

5.4 Comparing Modeling Performance

Similar to the analysis performed before, we use process models from the organizational handbook to compare the modeling performance of the t.BPM workshops with the established workshop productivity in the company.

We gathered data from a recently finished modeling project in which 18 processes from the e-mobility department were modeled. For that project we could determine the workshop time spent on each model. We calculated the productivity as the amount of information created per workshop time. For the tangible modeling workshop we did the same metric based on four snapshots of the evolving model in the tangible modeling workshop.

Figure 9 depicts the workshop productivity (y-axis) in relation to the size of the model (x-axis). In the four tangible modeling phases, 19–47 information nodes were mapped within 30–72 min. The average amount of information nodes added per hour varied from 25 to 39. In comparison, the software-supported workshop productivity varied from 6 to 67 nodes per hour.

In Fig. 9 we further see, that the t.BPM modeling productivity seems stable no matter the model size. In other words, the productivity does not change dramatically with larger models, in particular it does not decrease. Software-supported modeling workshops have a higher dynamic. They are less productive than t.BPM for models with less than 40 nodes. Their productivity increases dramatically when the model grows beyond 60 nodes.

We investigated the reason of this unexpected effect. We found copy-and-paste to be the dominant accelerator for digital modeling tools. The bigger the models, the

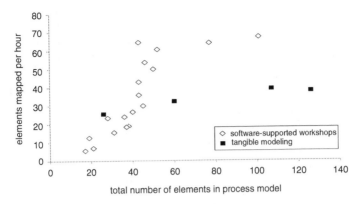

Fig. 9 The boxplot diagram shows the size of process models (x-axis) in relation to the productivity of the workshop in which they were created (y-axis) as the amount of elements mapped per hour. While t.BPM productivity is quite stable, software-supported workshops get more productive when the model grows big

more substructures get re-used. As an example, in accounting the invoicing is handled differently according to the contract type. After the process for one contract type is modeled, it gets copied and adapted for the other case. In the end, the different paths involve different IT systems but do not differ much in the steps taken or the roles responsible for the steps. By copying and pasting model parts, a digital modeling tool can speed up the creation of model information dramatically with up to 67 nodes modeled in 1 h in the sample project evaluated. This relation was an unexpected finding. We see model size and reusability of other model parts as a candidate to determine the tool choice for modeling workshops.

Our conclusion is that workshops are competitive in productivity if reuse of models is not possible, e.g. because a new area is to be modeled. We also point out that t.BPM modeling workshops require only one facilitator while software-supported modeling workshops require an additional software tool operator during the workshop.

6 Limitations of Research Results

Successfully applying t.BPM workshops has some context assumptions, such as a problem worth discussing among the people at the table. If assumptions about the context change, the results may not be the same and the findings presented then have to be re-evaluated. Given the same assumptions, there are still situational factors that are influential, such as the mood of the people and the empathy of the consultant. This is to say that the findings are situational and so should be the choice for or against t.BPM workshops.

Fig. 10 Snapshot from the application of t.BPM for process modeling training at REFA. Smaller groups model processes and share (see picture) their result with the other trainees

7 Further Fields of Application

We designed and evaluated t.BPM for process elicitation workshops moderated by a BPM consultant. Throughout our research, we also encountered further fields of application for t.BPM. In this section, we have shared just two situations in which the t.BPM can be applied. The dissertation on tangible business process modeling (Luebbe 2011) gives a complete overview about field-tested application scenarios for t.BPM.

Educating business process modeling experts. People who are educated in process modeling typically participate in process modeling trainings. The trainees have some introduction to process modeling by means of front lectures. The t.BPM modeling session is the very first hands-on modeling experience. They work out the process model in peer groups. The teacher walks around and may intervene to steer the modeling, but in general the participants should work out the model in a group discussion. People learn the skill of process modeling without having to learn the software tool first. The t.BPM tool is intuitive and therefore reduces the cognitive effort (Sweller 1988) of the learners.

For these workshops, we can reuse the t.BPM method knowledge such as the warm-up exercises. We tried this type of application in two cases. The situation in Fig. 10 shows a model evaluation discussion during the BPM training at REFA[2] in 2011.

Design thinkers ideating future services. Design Thinking problems to be solved in d.schools often require solutions to scenarios which play out 10 to 20 years from the current date. Discussing future services that can solve the problem requires a lot of envisioning. The t.BPM approach can provide a very concrete process-oriented viewpoint. Modeling the process raises issues regarding the value chain that enables the fulfillment of a new service. The d.school students

---

[2] http://www.refa-berlin.de/

Fig. 11 Snapshot from one of the d.school teams in Potsdam using t.BPM to model their envisioned future customer experience as a process using t.BPM

are from various backgrounds by nature. The t.BPM tool provides a basic framework of process concepts to talk about.

The d.school groups are not facilitated by a process modeling expert but get a short introduction to t.BPM and work it out from there by themselves. The objective of the session is to envision, share ideas, and build on the ideas of others. The modeling framework can be extended, for example with desires, pain points, or flying cars. The d.school students are not bound to formal process modeling. Instead, they are expected to jump out of the frame and add new aspects to the process model.

We prototyped this type of application with ME310 students at University St. Gallen, the global team-based design innovation class,[3] and with students at the HPI school of Design Thinking in Potsdam.[4] Figure 11 depicts a modeling situation at the d.school Potsdam. The students modeled their envisioned solution as a process from the user experience point of view.

8 Conclusion

This chapter reports on the third year of our research work related to t.BPM. The first year was dedicated to prototyping and tool development. In the second year, we evaluated the effect of our solution, the t.BPM toolkit, on individuals in laboratory environments. The results of those phases can be found as chapters in previous volumes of this book series. In 2011, we concentrated on field applications for tangible modeling.

---

[3] http://me310.stanford.edu/

[4] http://www.hpi.uni-potsdam.de/d_school/home.html

We joined hands with professional BPM consultants to run workshops in real BPM projects and to develop best practices for the application of t.BPM. The result is a set of method cards for BPM experts who want to setup and run t.BPM workshops. Additionally, we characterize process models that can benefit from t.BPM workshops, such as radical redesign projects. The results have been iterated with BPM consultants and researchers.

We measured and compared the performance and results of t.BPM workshops. We found that t.BPM is competitive with software-supported workshops for specific situations. If models get larger they may be created faster using copy-and-paste in a software tool. But t.BPM is stable in productivity and comparably fast for models that cannot gain speed boost through copy-and-paste. That is in line with our characterization that proposes using t.BPM in untapped process areas and for radical redesign rather than revision of well-established processes.

Future research. We found t.BPM to be a reasonably mature workshop technique by comparing it to other types of process elicitation. While there is a large body of comparative research on general requirements elicitation, see Davis et al. (2006) for examples, there is little comparative research specific to process elicitation techniques, with Rittgen (2009) as an exception. By comparing the different process existing elicitation techniques in a standardized way, future research could contribute to a deeper understanding about the best situational technique to choose from.

We opt to look into another stream of research: the sharing of information with clients. We showed with t.BPM that it is possible to strongly involve domain experts in the creation of models – here process models – that are relevant to software engineering. We think this idea of intuitive tooling and tangible interactions with clients can be beneficial for more aspects of the software requirements engineering project. We will further investigate information relevant to software engineering that could be created by domain experts using appropriate facilitation.

Acknowledgments We thank Markus Guentert, the student who conducted parts of the t.BPM field studies, for his support in this research. Moreover, we are grateful to the brave men and women who tried out t.BPM in their work environment. In particular, we'd like to thank Rüdiger Molle and Claas Fischer. They were the BPM consultants confident enough to apply t.BPM in real-world projects with their clients and allowed us to observe them.

References

Baskerville RL (1997) Distinguishing action research from participative case studies. J Syst Inf Technol 1(1):S.24–S.43

Baskerville RL (1999) Investigating information systems with action research. Commun AIS 2 (3es):S.4

Brydon-Miller M, Greenwood D, Maguire P (2003) Why action research? Action Res J 1(1):S.9

Davis A et al (2006) Effectiveness of requirements elicitation techniques: empirical results derived from a systematic review. In: Proceedings of the 14th IEEE International Requirements Engineering Conference, Minneapolis, pp S.179–S.188

Davison R, Martinsons MG, Kock N (2004) Principles of canonical action research. Info Syst J 14(1):S.65–S.86

Edelman J (2009) Hidden in plain sight: affordances of shared models in team based design. In: Proceedings of the 17th international conference on engineering design, ICED'09, Stanford, CA, USA, pp S.395–S.406

Edelman J (2011) Understanding radical breaks: media and behavior in small teams engaged in redesign scenarios. Dissertation. Stanford University

Edelman J, Grosskopf A, Weske M (2009) Tangible business process modeling: a new approach. In: Proceedings of the 17th international conference on engineering design, ICED'09, Stanford, CA, USA

Field AP (2009) Discovering statistics using SPSS. SAGE publications Ltd, London

Grosskopf A, Edelman J, Weske M (2009) Tangible business process modeling – methodology and experiment design. In: Proceedings of the 1st international workshop on empirical research in business process management (ER-BPM'09). Springer, Ulm, pp S.53–S.64

Keller G, Nüttgens M, Scheer AW (1992) Semantische prozessmodellierung auf der grundlage "ereignisgesteuerter prozessketten (epk)". Veröffentlichungen des Instituts für Wirtschaftsinformatik 89

Lewin K (1946) Action research and minority problems. J Soc Issues 2(4):S.34–S.46

Luebbe A (2011) Tangible business process modeling – design and evaluation of a process model elicitation technique. Dissertation, Hasso Plattner Institute for IT Systems Engineering, University of Potsdam

Luebbe A, Weske M (2010) Designing a tangible approach to business process modeling. http://ecdtr.hpi-web.de/report/2010/003/

Luebbe A, Weske M (2011a) Investigating process elicitation workshops using action research. In: Proceedings of the 2nd international workshop on empirical research in business process management (ER-BPM'11), Clermont-Ferrand, France

Luebbe A, Weske M (2011b) Tangible media in process modeling – a controlled experiment. In: Proceedings of the 23th conference on advanced information systems engineering (CAiSE 2011), London, United Kingdom, pp S.283–S.298

Object Management Group (2011) Business process model and notation (BPMN) 2.0

Plattner H, Meinel C, Leifer LJ (eds) (2010) Design thinking – understand, improve, apply. Springer, Berlin/Heidelberg

Plattner H, Meinel C, Leifer LJ (eds) (2011) Design thinking – studying co-creation in practice. Springer, Berlin/Heidelberg

Rittgen P (2009) Collaborative modeling of business processes: a comparative case study. In: Proceedings of the 2009 ACM symposium on applied computing, Honolulu, Hawaii, USA, pp S.225–S.230

Rittgen P (2010) Success factors of e-collaboration in business process modeling. In: Proceedings of the 22th conference on advanced information systems engineering (CAISE 2010), Hammamet, Tunisia, pp S.24–S.37

Susman GI, Evered RD (1978) An assessment of the scientific merits of action research. Adm Sci Q 23(4):S.582–S.603

Sweller J (1988) Cognitive load during problem solving: effects on learning. Cogn Sci 12(2): S.257–S.285

Towards a Shared Repository for Patterns in Virtual Team Collaboration

Thomas Kowark, Philipp Dobrigkeit, and Alexander Zeier

Abstract With *analyzeD* we established a platform that provides "out-of-the-box" monitoring capabilities for virtual team environments and enables the sharing and evaluation of recorded collaboration activities within a larger research community. Building on lessons learned from previous applications, we now present a refined and extended version of this platform. Its core feature is the possibility to share abstracted parts of the collected team collaboration networks with other users of the platform and, thus, broaden the basis for the validation of their influence on team performance. By that, these network sequences might elevate from being just coincidently reoccurring collaboration behavior to collaboration patterns (or anti-patterns), whose occurrences are strong indicators for the possible success of teams.

1 Introduction

The ubiquity of virtual collaboration in today's working environments is both a blessing and a curse. Teams are able to share information, communicate with each other, and coordinate their work through a wide variety of channels (Maznevski et al. 2000). They can also work in globally distributed teams without having to face many of the impediments that such settings used to have just a couple of years ago. However, with the growth of virtual collaboration, it is becoming increasingly difficult to keep track of all ongoing conversations, maintain an overview of all available information, and detect potentially detrimental collaboration behavior.

With *analyzeD* (Kowark and Uflacker 2011), we have created a platform that targets providing a simple and readily accessible way of collecting and analyzing information about artifacts of virtual team collaboration (e.g., wiki pages, email, shared folders, etc.). By combining them into singular graphs, so-called Team

T. Kowark • P. Dobrigkeit • A. Zeier (✉)
Hasso Plattner Institute, University of Potsdam, 14482 Potsdam, Germany
e-mail: alexander.zeier@hpi.uni-potsdam.de

H. Plattner et al. (eds.), *Design Thinking Research*, Understanding Innovation,
DOI 10.1007/978-3-642-31991-4_13, © Springer-Verlag Berlin Heidelberg 2012

Collaboration Networks (TCN) (Uflacker and Zeier 2010), our approach allows not only limiting the analysis to a single collaboration medium, but also incorporating data from multiple sources and the relations between them (e.g., emails containing links to wiki pages, wiki pages linking back to files, etc.). Furthermore, the software-as-a-service (SaaS) nature of the solution allows us to build up a database of TCNs that can be used to verify assumptions about the effects of certain team collaboration behavior with sample set sizes, which otherwise would be unattainable for a single researcher or even institutions.

Though this platform and its predecessor – the *d.store* (Uflacker 2009) – were successfully applied in the analysis of multiple projects in the domains of Engineering Design (Skogstad 2009; Uflacker 2010) and Software Engineering (Kowark et al. 2011), it became apparent that further development efforts were necessary. Firstly, the SaaS concept of the platform required a revision. While it would be highly beneficial from a research point of view to have all data collected within a single database, data privacy concerns quickly became impediments for real-world adoption; especially by companies. Secondly, the platform was previously lacking the means to share analysis results in a structured manner.

In this chapter, we outline modifications and extensions of our system that tackle the aforementioned issues. The revised architecture aids data privacy as it ensures that the original TCNs remain in possession of their creators and only basic information about the networks is transferred to a central service. Subsequently, we introduce the notion of collaboration patterns as a means to share results of TCN analysis with other users of the platform. In the final section, we use the example of two engineering design projects to highlight how defining collaboration patterns on the level of TCN can help to detect different virtual collaboration behavior of otherwise seemingly equal teams. The chapter closes with an outline of next steps in our research.

2   analyzeD: A Shared Platform for Virtual Team Collaboration Research

The *analyzeD* platform was originally designed as a software-as-a-service (SaaS) version of its predecessor – the *d.store*. The d.store is a customizable service platform that enables users to collect and analyze virtual collaboration activities in a non-interfering manner during project runtime. It is configurable and can be utilized to capture collaboration activities from heterogeneous groupware systems, generating a single record of temporal and semantic relationships between identified actors and resources.

As a research prototype, this platform was not optimized for user experience. Setting up the platform, as well as the accompanying services for capturing collaboration events from the different data sources is a technically challenging task. With analyzeD, we wanted to make the capabilities of the d.store available to a

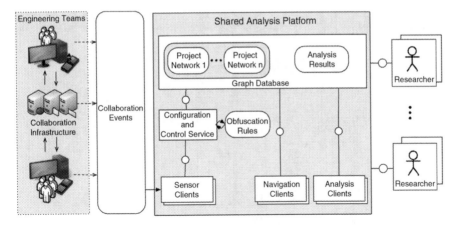

Fig. 1 Basic architecture of the analyzeD platform

wider audience. Therefore, simplicity became the main focus: the system was supposed to be easy to configure, provide a simple interface for data collection, and powerful analysis methods that could be used without in-depth technical skills.

2.1 Architecture and Standard Workflow

Figure 1 presents the architecture of the first version of the analyzeD platform. The building blocks of the system reflect the previously determined requirements. In the following, we briefly outline which parts of the system are responsible for the provision of certain functionality. To this end, we provide a stepwise introduction of the standard use-case for the system.

2.1.1 Data Collection

Due to the SaaS nature of the solution, registration is mandatory before being able to use the system. After that, users are able to create new *Projects*. Each project requires some basic information, such as project domain, start and end date, a unique name, and a brief description that would allow other users to reason about the background of the project. Additionally, users have to choose which data sources should be used for the analysis. The system provides ontologies for a variety of standard collaboration artifacts, such as wikis, emails, shared web folders, version control systems, and issue tracking systems. For each of the selected ontologies, the user can further choose, which particular systems need to be parsed.

This choice selects the *Sensor Clients* that will be used to collect data. Each sensor client is specifically tailored to parse information from a distinct data source

Fig. 2 analyzeD interface for collaboration data collection

and translate this information into a format that can be handled by the analyzeD platform. New sensor clients, e.g., for a newly created collaboration tool, have to implement a well-defined interface in order to be recognized by analyzeD. Once registered with the platform, all platform users can access these services.

After the project is set up, the users simply have to provide valid access credentials for the respective sensor clients and the system is ready to collect the data from the desired sources (see Fig. 2). All data will be uploaded to the platform and combined into a singular TCN for the project.

2.1.2 Data Analysis

Once the data is stored within the TCN, it is possible to analyze the reflected collaboration behavior. The main interface for data analysis is a SPARQL (SPARQL Protocol and RDF Query Language) (Prudhommeaux 2008) endpoint. This language allows detecting sets of triples within the graphs that fulfill the constraints specified in the query. During some minor sample applications, it became apparent that an abstraction of this language was necessary to foster widespread usage of the platform. Hence, we created a solution that allows the creation of SPARQL queries through a graphical interface (see Fig. 3).

The "What?" element on the left allows choosing a class that is of interest to the user. Depending on the chosen class, the "Attribute?" select list is populated with all attributes and relations that are known for this particular class. In the third step, a custom input depending on the type of attribute or relation is displayed. So, depending on whether an attribute is of type String, Integer, Date, etc., the users have different input methods (compare lines 1 and 3). If users choose a relation (i.e., an attribute whose value is not a literal value but a reference to another node of the TCN), they subsequently can specify constraints about the entity the relation is

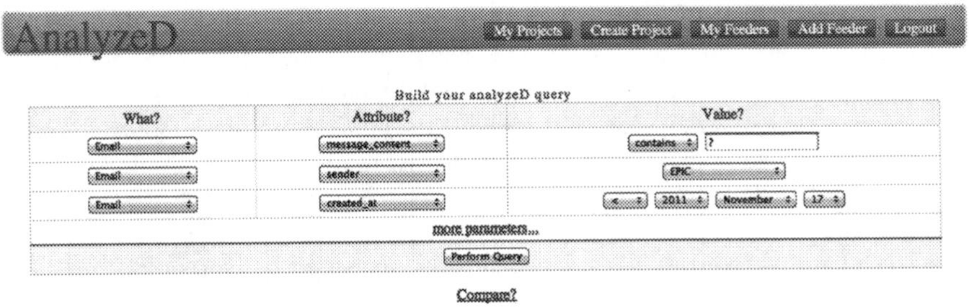

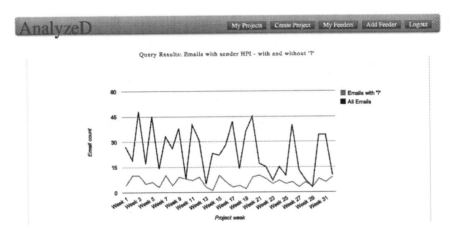

Fig. 3 Guided SPARQL query creation

Fig. 4 Diagram presenting the result of a SPARQL query

pointing to. The query depicted in Fig. 3, for example, will return all *Emails* that contain a question mark, have been sent by a person of type "EPIC" and were written before November 17, 2011. Furthermore, a compare button is available that allows users to specify a second query (e.g., for senders of type "Stanford") to compare collaboration behavior exposed by people from different institutions. The result of the query can be presented either in textual form or as simple diagrams (see Fig. 4).

2.1.3 Sharing of Collaboration Data and Analysis Results

The intention of TCN analysis is the detection of possibly beneficial or detrimental collaboration behavior. The presented query interface might be an aid in that regard, but the most important feature of the platform is the possibility to share collaboration data and analysis results with other users.

Due to the SaaS nature of the system, the implementation of sharing collaboration data is straightforward. Each created TCN is stored within the central database and other users can access this information to verify their hypotheses about the impact of certain behavior. To ensure data privacy, only the original creator of the data is able to see the real names of people and contents of, for example, emails and wiki pages. Other users are only able to see obfuscated variants of this information. Beyond exploring other projects, it is possible to choose a sample set for analysis by specifying certain key parameters. Hence, SPARQL queries cannot only be issued against single TCNs, but the users can choose to perform them on any TCN matching the given criteria. These criteria, for example, include size of the network, duration of the project, or the domain it took place in. The possibility to share analysis results is presented in detail in next section.

2.2 Architecture Adjustments

The described version of the system is currently deployed at our institute and used in multiple projects to analyze the collaboration behavior of students teams both ex post facto and in real-time while the projects are still running. In this setting, the system works just fine. Researchers from collaborating universities have obtained access credentials as well and are beginning to use the system for their work. But, the centralized nature of the service has its downsides.

We received requests for using the system, but after giving some demonstrations, the requestors ultimately shied away from making a decision. The main reason: data privacy issues. The basic obfuscation functionality offered by the original system was deemed insufficient for their purposes, as data could only have been stored at our servers if it was obfuscated to a point where it would vastly affect the analysis process. Hence, the need for storing the collaboration traces at the respective companies or universities emerged and was reflected by a change in the platform architecture (see Fig. 5).

Instead of one central instance, companies and universities can obtain their own instance of the system and either deploy it on a dedicated server or simply run it inside a provided virtual machine. By that, we are able to retain the ease-of-use of the original platform, since the only additional step would be downloading and starting the virtual machine.

Maintaining the notion of a *shared* platform, however, required some additional effort. Instead of sharing all collaboration networks at a central point, the new version of the platform only contains a central repository of analysis results. In addition to the definition of the results, which is presented in the next section, this repository also maintains statistics about their occurrences (i.e., how often does certain behavior occur in all networks that the repository knows of). Therefore, the central repository still needs access to the distributed instances of the analysis platform, but instead of the real data, only basic information about the stored collaboration networks is transferred.

With the help of these adoptions in platform design, we were able to maintain the basic principles of the original analyzeD concept, and at the same time respect data privacy issues. The new systems leaves the original creators in possession of the

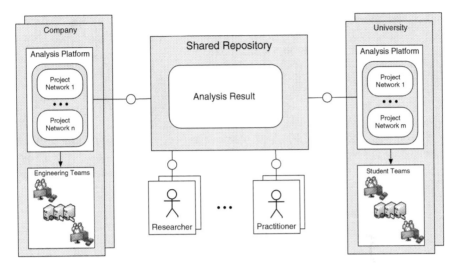

Fig. 5 Architecture of the distributed platform with a central pattern repository

team collaboration networks, yet still enables researchers and practitioners to share insights about virtual team collaboration with one another and verify their findings against all networks known to the platform.

3 Patterns in Virtual Team Collaboration

A key concept of the previous section is the "sharing of analysis results". These results usually refer to some parts of the analyzed team collaboration networks that have an impact on the performance of the teams under investigation. In this section we introduce the notion of collaboration patterns as a means to define such behavior and share it with other users of the analyzeD system. Our approach is based on the definition proposed by Dai et al. (2009):

> A collaboration pattern is a prescription, which addresses a collaborative problem that may occur repeatedly in the environment. It describes the forms of collaboration and the proven solutions to a collaboration problem and appears as a recurring group of actions that enable efficiency in both the communication and the implementation of a successful solution. The collaboration pattern can be used as is in the same application domain or it can be abstracted and used as a primitive building block beyond its original domain.

The TCN that are stored within our platform represent the concrete sequence of collaboration events that occurred within a project team. Collaboration patterns, thus, are generalized sub-graphs of such TCN. If, for example, sending emails with hyperlinks to previously edited wiki pages to other team members and superiors has proven to be beneficial within a project, the respective TCN contains multiple different traces that all represent this behavior but slightly differ from one another (e.g., different timespans between editing the page and sending the email, different page names, etc.). To adequately abstract such traces, five concepts needed to be covered by a definition of collaboration patterns:

- *Attributes* are necessary to define value constraints for the nodes that are part of each pattern.
- *Tags* allow the definition of custom classes that a node needs to have.
- *Sequences* define the order of events within the pattern.
- *Cardinality* allows specifying the number of people or collaboration artifacts involved in the activity.
- *Groups* denote that certain nodes within the pattern can be instances of a set of classes.

3.1 Formal Definition

For the sake of simplicity, the complete formal definition of collaboration patterns as directed, labeled, abstracted sub-graphs of TCN is omitted here. It is available for the interested reader as part of a separate publication (Kowark and Dobrigkeit 2011). In this chapter, we will instead concentrate on presenting the basic ideas with the help of an example pattern definition (see Fig. 6). It contains all of the aforementioned five concepts that need to be covered by the definition. The graphical notation was developed as an aid for the description of patterns, as it closely resembles the graph structure of the TCN and, thus, can be easily created by "cutting-out" parts of the original networks and abstracting from the actual attribute values and time constraints. Users can create these descriptions by using the *Visual Creator* depicted in Fig. 5.

Nodes are represented as circles that are labeled with possible classes of the node. The label can be extended by a set of valid tags for a node, which is depicted in squared brackets. Relations between nodes are represented by solid, directed edges. The label of an edge denotes the name of the relation and optionally defines cardinality constraints. These cardinalities define the minimum and maximum number of relations that need to be present in order to match the pattern. That means, in the example, the email requires a minimum of two receivers. If no cardinality is defined, at least one relation is expected to be present. Sequences can be expressed by dashed edges that connect two relations. Again, a minimum and maximum value can be specified, along with an abbreviation of the desired time unit (i.e., s = seconds, h = hours, d = days). Finally, attribute constraints are expressed within a separate rectangle that is attached to the respective node. They can be defined using SPARQL syntax and, thus, support all available features, such as regular expression matching, integer comparisons, or date constraints, among others. The example requires the label of each *WikiPage* or *Ticket* to contain the word "important".

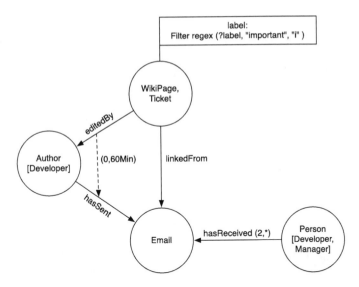

Fig. 6 Collaboration pattern example – sending emails with hyperlinks to important tickets or wiki pages to fellow developers and superiors within 60 min after editing the respective artifacts

3.2   Pattern Matching

To allow storage of patterns, an ontology was created using the Web Ontology Language (OWL) (McGuinness 2004). It uses concepts introduced in the definition of TCN to simplify the process of matching patterns within collaboration networks, as no further translation is required. Additionally, OWL allows for automatic checking of patterns for compliance with the ontologies used in a given TCN through reasoning. In order to detect pattern occurrences within TCN, a translation into SPARQL queries is necessary. The mapping is straightforward for nodes, relations, attributes, tags, and groups as they can be directly expressed using standard SPARQL constructs. However, date arithmetic functionality is not part of the query language and the proposed date and time logic of the patterns has to be added programmatically by scanning through the result set of the query and rejecting results that do not fulfill the constraints.

3.3   A Shared Pattern Repository

The formal definition of collaboration patterns and the mapping to SPARQL queries are prerequisites for the implementation of a shared pattern repository. Building on the presented concepts, a prototypical implementation was created. Foundation for the technical implementation is an ontology that translates the formal definition of the concept of collaboration patterns to RDF/OWL.

3.3.1 Pattern Instances

Each pattern that is defined by platform users is an instance of this concept. It consists of an internal Universal Resource Identifier (URI), the name of the pattern and an optional description. Furthermore, the actual nodes and relationships, which form the pattern instance, are defined and stored. The node and relationship types are directly derived from the concept ontologies that define the TCN to ensure compatibility between patterns and TCN.

3.3.2 Network Instance

A generic pattern definition can never incorporate specialties of each project. Certain aspects, such as attributes or tags, need to be instantiated on a network level (e.g., by translating tags specific to software engineering projects to the terminology used in an engineering design context). Additionally, one can change values for cardinality on network instances. For example, a project with five members cannot easily be compared to a project with 20. Thus, network instances offer the possibility to adjust cardinality constraints of patterns for each TCN.

3.3.3 Match Instances

Once patterns have been translated to match the peculiarities of specific TCN, matching can be performed. Each match represents a different instance of a pattern. Match instances are stored in the pattern catalog as well. They form the database to do further analysis and gather user feedback on the usefulness of a certain pattern. For every resource and relation of the matched instance in the network, a link to the instance of the pattern is stored.

3.3.4 Pattern Statistics

To verify whether certain patterns are actually beneficial or detrimental for project progress, large data samples need to be collected and analyzed. The implementation is capable of automatically collecting statistics about pattern matches for all TCNs available within the platform. The most interesting information is the number of match instances found across all TCN. This is an indicator for the generality of a pattern, i.e., whether the depicted collaboration behavior is a common sequence of actions or specific to a particular network.

Given the number of occurrences of a certain pattern within a network, a statistical analysis over all networks can be conducted. However, since not all networks are comparable, the average has to be put into perspective. Possible quantifiers are the runtime of the project (occurrences of a pattern per time unit) or size of the network in terms of number of nodes. The total run-time of the events in a TCN can be calculated based on the time information that is associated with each node and relationship.

3.3.5 Pattern Rating

Collaboration patterns alone cannot capture the full complexity of team behavior. Additional manual evaluation and input is required to build a set of patterns, which can provide the best practice for teamwork behavior. To aid this process, the repository provides features to gather feedback and rate each match instance of a pattern in various categories:

- *Participation*: Was the user directly involved in the pattern instance, for example as the recipient of an email, or author of a wiki page?
- *Expected Context*: Each pattern is designed to describe behavior that occurs under certain circumstances. Thus, an expected result can be stated beforehand. When rating a pattern instance, the user should give feedback, whether the actual context relates to the expected context. For example in concurrent conversations (i.e., users simultaneously respond to different replies in a multi-user email conversation), it is expected that, sometimes, important information is lost because replies to e-mails are overlooked. Given the actual instance of the pattern, the user is able to analyze the content of the emails and state whether the reply contained information that was overlooked or not.
- *Subjective Usefulness*: Did the user experience the actions in the pattern as helpful or hindering in respect to the progress of the project? Given enough data samples, this attribute might allow converting patterns into best practices for team collaboration behavior.

Figure 7 presents a screenshot of a prototypical implementation of the repository. Patterns are categorized to simplify the analysis of different aspects of virtual team collaboration.

In this section, we presented the notion of collaboration patterns as a means to share the results of virtual collaboration analysis with other platform users. Desirable or unwanted behavior can be modeled by using a simple graphical notation and is instantly available in a shared pattern repository. Consequently, all TCNs can be analyzed for occurrences of this pattern to further strengthen or weaken hypotheses about the impact of the pattern on team performance.

4 Example Application: Modeling Collaboration Behavior of Student Project Teams

To test the modeling of patterns and the detection of matching instances in TCN, an analysis was conducted using data from a global project-based engineering design course. This course has a long-standing history of combining teaching with research and has served as a data source for various papers and dissertations. In the following, we present an excerpt from a comprehensive analysis of collaboration behavior in two course projects. For further examples, please refer to the original publication (Dobrigkeit 2011).

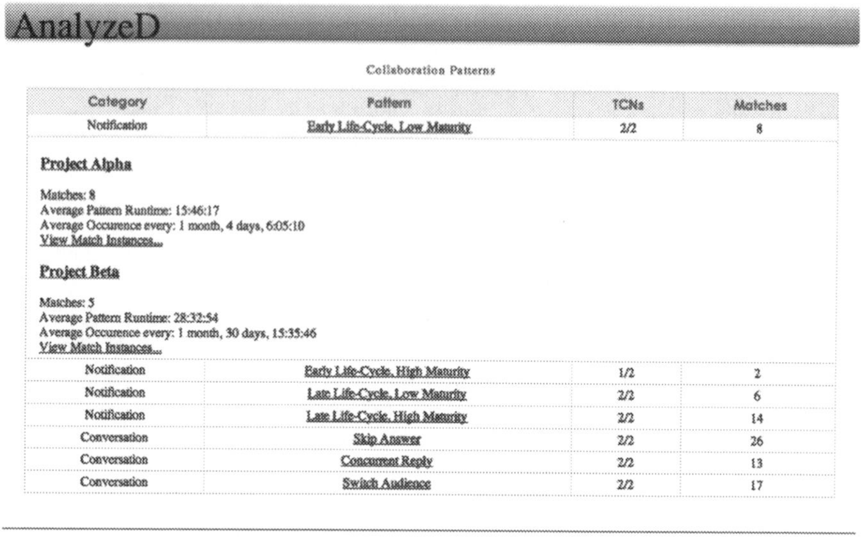

AnalyzeD

Collaboration Patterns

Category	Pattern	TCNs	Matches
Notification	Early Life-Cycle, Low Maturity	2/2	8

Project Alpha

Matches: 8
Average Pattern Runtime: 15:46:17
Average Occurence every: 1 month, 4 days, 6:05:10
View Match Instances...

Project Beta

Matches: 5
Average Pattern Runtime: 28:32:54
Average Occurence every: 1 month, 30 days, 15:35:46
View Match Instances...

Notification	Early Life-Cycle, High Maturity	1/2	2
Notification	Late Life-Cycle, Low Maturity	2/2	6
Notification	Late Life-Cycle, High Maturity	2/2	14
Conversation	Skip Answer	2/2	26
Conversation	Concurrent Reply	2/2	13
Conversation	Switch Audience	2/2	17

Fig. 7 Pattern repository overview page

4.1 Data Collection

The data analyzed in this section originates from two projects (in the following referred to as *Project Alpha* and *Project Beta*). Teams of six to eight students from two different universities are given a design task by a corporate partner. The two teams are separated over different continents and have to deal with the complexity of working globally and in different time zones. Each project has several stakeholders the students have two address. From the corporate partner at least one liaison is the primary contact for the team. From the universities each sub team has a teaching staff they can ask for help and guidance. Additionally, a coach is assigned to each sub team that is not involved in the grading process. The coach can provide feedback as well and be a helpful neutral contact in case of problems between the teams and the stakeholders. Both the liaisons and the teaching staff provide regular feedback on the progress of the project.

Over the course of 9 months the teams have to organize their work, develop ideas and finally build working prototypes of their product. The course is split up into roughly three different phases, which require very different communication and collaboration techniques from the students. The analysis of the projects has been done throughout a single class of the project and while all the teams had different design tasks, the course framework makes the projects comparable. More information on the background of the class can be found in Carleton (2009).

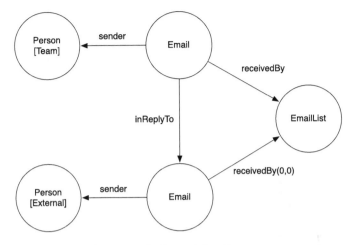

Fig. 8 External communication pattern for forwarded replies

4.2   Analyzing External Communication

The characteristics of the communication with external contacts of the teams, for example, their corporate liaisons or experts that are not part of the teaching team, provide indicators for project success in the context of this course (Uflacker and Skogstad 2009). In the following, we present two varieties of this communication behavior and outline how modeling them as collaboration patterns and detecting the occurrences of these patterns within the collaboration networks can be used to reveal the different working styles of the teams under investigation.

4.2.1   Forwarding Communication Pattern

A single team member contacts the external person via e-mail from his personal e-mail account. As these accounts are not monitored, this conversation cannot be analyzed, but at a certain point the team member might decide that the conversation contains important information that needs to be shared with the whole team by forwarding an email from the external contact to the mailing list. This type of communication happens, for example, when first approaching a new contact or scheduling a meeting with the external contact and afterwards informing the team about the date, time, and location. As these forwards are mainly informational it is expected that the number of replies be below average. The corresponding pattern is shown in Fig. 8.

Table 1 shows the averages of replies for this type of e-mail. Direct replies refer to e-mails that reply directly to the original message. Total replies additionally include any further follow-up replies. When comparing the numbers for all e-mails with the numbers for forwarded messages the difference becomes clear. For both

Table 1 External communication pattern for forwarded replies

	Total	Direct replies	Total replies
Project Alpha			
All e-mails	910	0.43	0.73
Forward	47	0.32	0.57
External reply	154	0.35	0.63
Project Beta			
All e-mails	1,512	0.55	1.31
Forward	63	0.21	0.62
External reply	264	0.51	1.14

projects the numbers are well below average. For Project Alpha, the team with fewer total e-mails, the numbers are closer to the average, but for Project Beta a forwarded e-mail received fewer than half the replies of an average e-mail.

4.2.2 Team Communication Pattern

The second pattern related to external contacts occurs when the original e-mail was sent in cc to a project mailing list and the external contact replies directly to the whole team in the form of the mailing list. The original e-mail is often written as coming from the whole team, and often to people closely involved with the project or already familiar with the team members, mainly the corporate liaison. The graphical representation of this pattern is depicted in Fig. 9.

While the expectation is that these e-mails actually would receive more replies than an average e-mail, the numbers from both projects do not support this hypothesis. Table 1 shows the number for external replies to the mailing lists. While being closer to the total average than forwarded e-mails, the numbers are still below the total average. A possible explanation could be, that both teams are more likely to change the communication channel from e-mail to, for example, phone or videoconference, when in contact with external partners.

The two examples presented in this section highlight how the previously introduced notion of collaboration patterns can be used to model communication behavior that potentially has an impact on team performance. The two patterns presented in this section are actually stored within the shared pattern repository shown in Fig. 5 and, thus, the communication of future projects can be monitored for their occurrence. By that, the teaching team is relieved from constantly keeping track of the project mailing lists by themselves and will conveniently be notified once a new instance of either of the patterns is detected. Subsequently, they can then analyze this communication manually and provide focused feedback to the student teams.

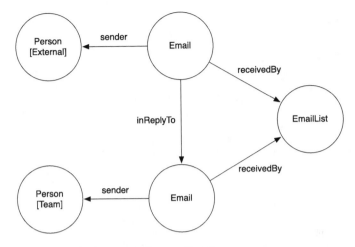

Fig. 9 External communication pattern for e-mail replies

5 Conclusion and Outlook

"We are drowning in information but starved for knowledge" (J. Naisbitt). This quote is one of the key drivers for the development of the analyzeD system. As soon as it was possible to easily create single graphs that contain all available information about collaboration events happening in the multitude of systems that support today's engineering teams, it became apparent that the amount of data is constantly becoming harder to handle and that asking the right questions about the data is the real problem at hand.

This problem was tackled from two sides. Firstly, we created a platform that allows us to explore the given data by means of a guided query interface and thus help researchers that might know the right questions but lack the capability to formulate them in a technical query language. Secondly, we enabled said users to share the results of their analysis with other users of the platform that might have the technical capabilities but do not know the right questions to ask. By bridging the gap between these two worlds, the system might contribute to a better understanding of virtual team collaboration processes.

Technology alone, however, is not sufficient to create a common service for capturing and analyzing virtual collaboration activities that promotes comparative research and team diagnostics in engineering design. Without a widespread usage of the system and scientific proof of its benefits, it is unlikely that the desired impact can be achieved. Therefore, future work will concentrate on these aspects. A controlled experiment is currently being conceived to validate the impact of shared analysis results on the ability to detect beneficial or detrimental behavior within large team collaboration networks. Furthermore, we are currently preparing the system for publication under an open-source license, which could improve its distribution throughout the research community and allow other developers to participate in the evolution of the system.

References

Carleton T, Leifer L (2009) Stanford's ME310 course as an evolution of engineering design. In: Proceedings of the 19th CIRP design conference-competitive design, Cranfield University, 30–31 Mar 2009

Dai X, Popplewell K, Wulan M, Apostolou D, Mentzas G, Papageorgiou N, Verginiadis Y, Das B, Harding J, Palmer C, Swarnkar R, Carpentier M, Lorre JP, Rajsiri V (2009) Collaboration patterns model and ontology. Technical report, ICCS, CU, LU, EBM

Dobrigkeit P (2011) Patterns in virtual design team collaboration. Master's thesis, Hasso-Plattner-Institut für Softwaresystemtechnik, Universität Potsdam

Kowark T, Dobrigkeit P, Zeier A (2011a) Towards a shared repository for patterns in virtual team collaboration. In: 5th international conference on new trends in information science and service science, October 2011

Kowark T, Müller J, Müller S, Zeier A (2011b) An educational testbed for the computational analysis of collaboration in early stages of software development processes. In: Proceedings of the 44th Hawaii international conference on system sciences (HICSS), January 2011

Kowark T, Uflacker M, Zeier A (2011c) Towards a shared platform for virtual collaboration analysis. In: The 2011 international conference on software engineering research and practice (SERP'11), July 2011

Maznevski M, Chudoba K (2000) Bridging space over time: global virtual team dynamics and effectiveness. Organ Sci 11(5):473–492

McGuinness DL, Van Harmelen F et al (2004) OWL web ontology language overview. W3C recommendation, 10

Prud'hommeaux E, Seaborne A (2008) SPARQL query language for RDF, W3C recommendation. http://www.w3.org/TR/rdf-sparql-query/

Skogstad P (2009) A unified innovation process model for engineering designers and managers. Ph.D. thesis, Stanford University, Center for Design Research

Uflacker M (2010) Monitoring virtual team collaboration: methods, applications, and experiences in engineering design. Ph.D. thesis, Hasso Plattner Institute for IT Systems Engineering, Potsdam

Uflacker M, Zeier A (2009) A platform for the temporal evaluation of team communication in distributed design environments. In: CSCWD'09: Proceedings of the 2009 13th international conference on computer supported cooperative work in design, Washington, pp 338–343

Uflacker M, Zeier A (2010) A semantic network approach to analyzing virtual team interactions in the early stages of conceptual design. Future Gener Comput Syst 27(1):88–99

Uflacker M, Skogstad P, Zeier A, Leifer L (2009) Analysis of virtual design collaboration with team communication networks. In: 17th international conference on engineering design (ICED'09), August 2009

Adopting Design Practices for Programming

Bastian Steinert, Marcel Taeumel, Damien Cassou, and Robert Hirschfeld

Abstract Developers continuously design their programs. For example, developers strive for simplicity and consistency in their constructions like practitioners in most design fields. A simple program design supports working on current and future development tasks. While many problems addressed by developers have characteristics similar to design problems, developers typically do not use principles and practices dedicated to such problems. In this chapter we report on the adoption of design practices for programming. First, we propose a new concept for integrated programming environments that encourages developers to work with concrete representations of abstract thoughts within a flexible canvas. Second, we present continuous versioning as our approach to support the need for withdrawing changes during program design activities.

1 Introduction

According to Herbert Simon, design should be considered a meta-discipline:

> Everyone designs who devises courses of action aimed at changing existing situations into preferred ones. The intellectual activity that produces material artifacts is no different fundamentally from the one that prescribes remedies for a sick patient or the one that devises a new sales plan for a company or a social welfare policy for a state ... (Simon 1996, p. 130)

Programming arguably involves design in various respects. In particular, developers design for end-users but also for developers. They continuously prepare their code base for current and future coding activities. Developers thus conduct intellectual activities that are similar to the activities of designers in other fields.

B. Steinert • M. Taeumel • D. Cassou • R. Hirschfeld (✉)
Software Architecture Group, Hasso Plattner Institute, University of Potsdam, Potsdam, Germany
e-mail: robert.hirschfeld@hpi.uni-potsdam.de

H. Plattner et al. (eds.), *Design Thinking Research*, Understanding Innovation,
DOI 10.1007/978-3-642-31991-4_14, © Springer-Verlag Berlin Heidelberg 2012

Also, the general understanding of programming matches well to the general descriptions of the design concept.

> Design as a noun informally refers to a plan or convention for the construction of an object or a system [...] while "to design" (verb) refers to making this plan. [...] However, one can also design by directly constructing an object (as in pottery, engineering, [...], graphic design) (Wikipedia: the free encyclopedia 2011)

In this sense, developers design by constructing the program. Unfortunately, there is no well-accepted formal definition of the design concept, for example, measured in terms of citation. However, the following recent attempt allows for emphasizing its connection to programming. (Concepts related to programming are inserted in angle brackets.)

> (noun) a specification of an object <the program>, manifested by an agent <the developer>, intended to accomplish goals, in a particular environment, using a set of primitive components <programming language, libraries, ...>, satisfying a set of requirements, subject to constraints <readability, performance, ...>; (verb, transitive) to create a design, in an environment (where the designer operates) (Ralph and Wand 2009)

Experience teaches us to approach similar problems in similar ways. What would it mean for a developer to work like a designer? While many problems addressed by developers have characteristics similar to design problems, developers typically do not use methods developed to explicitly address these problems. Divergent and convergent thinking, externalizing thoughts, supporting thinking by doing, working on parallel lines of thought—we believe that such design methods can significantly improve program design outcomes, which are increasingly important for software development and evolution.

In this chapter, we report on our ongoing work of adopting design practices to programming. After introducing the reader to program design—what is designed and for whom—we present two results of our research efforts that support developers in conducting program design activities. First, we describe a new concept for programming environments that encourages developers to work with concrete representations of abstract thoughts within a flexible canvas. Second, we will look at how prototyping activities typically interleave with the advancement of a program. This raises the need for going back to previous states of work. We will describe the lack of proper support and propose our approach to meet this important need.

2 Program Design

Developers continuously redesign their programs. A canonical reference about best practices in programming starts by listing some crucial questions developers face everyday (Beck 1997):

- How do you choose names for objects, variables, and methods?
- How do you break up the logic into methods?
- How do you communicate most clearly through your code?

In this section, we briefly introduce the need for program design explaining who is targeted during design activities and exemplifying some aspects of program design that are to be considered.

2.1 Design for Whom?: End-Users Versus Developers

By definition, design activities are targeted to the needs of users, having the goal of changing existing situations into preferred ones. According to this understanding, it is meaningful to distinguish between two kinds of users that are the target of design activities during software development:

- **User-experience design** targets end-users and involves the graphical layouts and workflows that guide users;
- **Program design** targets developers and involves meaningful names, indentation and code layout, abstractions, and separation into modules.

This report mainly deals with *design* activities that target developers.

2.2 Proper Arrangement and Meaningful Names

We will now look at the need for meaningful names and abstractions using a typical example.

Goal 1. Given a list of numbers: 3, 6, 1, 9 10, . . .
   Find all numbers smaller than 5.
   A possible way to fulfill this goal is by writing the following piece of Smalltalk code:

```
s := "... a list of numbers ..."
r := OrderedCollection new.
1 to: s size
do: [:i |
e := s at: i.
e < 5 ifTrue:
[r add: e]]
```

This piece of code basically creates a container (called r), iterates over the list of provided numbers, and inserts each number satisfying the criteria into the container.

The next piece of code presents exactly the same behavior but is non-arguably easier to read. First, names can increase the readability of a program if they clearly express the role of the constructs identified. Second, indentation of lines better conveys the structure of the source code:

```
givenNumbers := "... a list of numbers ..."
result := OrderedCollection new.
```

```
1 to: givenNumbers size do: [:i | | eachNumber |
    eachNumber := givenNumbers at: i.
    eachNumber < 5
        ifTrue: [result add: eachNumber]]
```

Such a readable code also helps other developers who might need to read and understand the code long after it has been written. Consider, for example, the following goal, which is very similar to the one stated above:

Goal 2. Given a list of names: Jones, Graham, Steele, ...
Find all names matching 'jones'.

Since the code above is well written, it is easy to identify the code fragments that need to be adapted for the new goal. The new goal can now be fulfilled by the following piece of code:

```
givenNames := "... a list of names ..."
result := OrderedCollection new.
1 to: givenNames size do: [:i | | eachName |
        eachName := givenNames at: i.
    (eachName matches: 'jones')
        ifTrue: [result add: eachName ]]
```

2.3   The Invention of Abstractions

Let's assume there is a third goal that is similar to the ones above:

Goal 3. Given a list of dates: 09/09/1999, 11/11/2011 ...
Find all dates before today.

A developer could certainly copy and adapt the previous piece of code. However, it becomes clear that a pattern emerges from the previous pieces of code. Such kind of repetition often encourages developers to think about new abstractions that can express such a common need more precisely. The following source code abstracts over iterating elements and filtering according to a predicate.

```
Collection>>select: aBlock
    newCollection := OrderedCollection new.
    self do: [:each |
        (aBlock value: each)
            ifTrue: [newCollection add: each]].
    ↑ newCollection
```

While the code is arguably harder to read, it allows for fulfilling all (three) goals using less code and being more concise.

```
"Given a list of numbers: 3, 6, 1,  9 10... find all numbers
smaller than 5"
```

```
givenNumbers select: [:each | each < 5]
"Given a list of names: Jones, Graham, Steele...
find all names matching 'jones'"
givenNames select: [:each | each matches:'jones']
"Given a list of dates: 09/09/1999, 11/11/2011...
find all dates before today"
givenDates select: [:each | each isBefore: Date today]
```

This new abstraction increases the expressiveness of the previous code snippets and therefore makes them easier to read and follow, given the reader understands the new abstraction.

2.4 Conceptual Models

Designing programs refers to the intentional effort of structuring an application's code base beyond fulfilling the strict needs of correctness and completeness. Thereby, like practitioners in most design fields, developers strive for simplicity and consistency in their constructions, which seems reasonable because of the following:

- **Structure helps manage immense amounts of information**, which can easily add up to the size of thousands of books or even an entire library (Kay et al. 2010),
- **Modularization eases change**, which can reduce the number of elements to be understood and touched on in subsequent development steps,
- **Order and simplicity support thinking**, which seem like a natural desire in enabling one to see more clearly. The arrangement of elements partially determines our perception and thoughts. Conceptual models like those presented in Fig. 1 implicitly act as frames that we use to understand the problem and solution spaces. Moreover, simplicity and order arguably make us feel better, which in turn positively affects creative thinking (Ashby et al. 1999).

These aspects do not affect the correctness of a program, but they do render program design activities important; their outcomes determine how the program is perceived, analyzed, and processed, which in turn affects subsequent development steps and their success.

3 Interactive Access to Program Run-Time Information

In software development, program comprehension is an ongoing challenge that developers need to face day by day. While complex software systems are developed, more and more essential knowledge about the corresponding problem domain

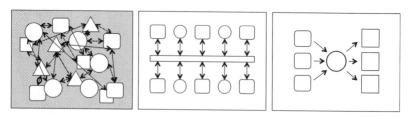

Fig. 1 Three different conceptual models of a program. The left model is less structured and exhibits less order than the other two, thereby rendering development tasks more difficult

is collected. A better understanding of the underlying problems leads to new or changed requirements, which themselves need to be fulfilled by developers. Generally speaking, software systems spend over 60 % of their lifetime within this iterative process of perfective and corrective maintenance (Meyer 2000). As periods of days, months, and even years pass by, chances decrease that developers write any new source code without having to consider formerly created work. Such existing work often raises challenges in understanding the underlying ideas and hence impedes developers' current maintenance tasks.

Developers rely on effective tool support to accomplish comprehension tasks (Sillito et al. 2008). Ideally, integrated programming environments should offer this support and enable them to fully understand highly dynamic run-time behavior of complex software systems. If developers understand such systems in-depth, they will maintain and extend them within shorter periods of time while making fewer mistakes. Unfortunately, existing tools provide only limited support for program comprehension activities.

In this section, we illustrate how developers should comprehend parts of existing software systems and explain which options they really have when using traditional environments such as Visual Studio.[1] Based on these insights, we describe the concept of a new programming environment that overcomes present limitations and strives for better program comprehension support.

3.1   The Need for Accessing and Arranging Run-Time Data

Developers access and process program-related information to understand abstractions that are described with programming language constructs. There are several sources of such information that differ in accessibility, actuality, and correctness:

- The *knowledge* about the source code in question should be possessed by its authors. Unfortunately, this knowledge may not be accessible because the person

---

[1] http://www.microsoft.com/visualstudio

is unavailable, is not allowed to expose details, or has already forgotten important aspects.

- External *documentation* provides a high-level source of information about basic ideas and guidelines for the system. This includes end-user or developer manuals that are supported with illustrative examples and code snippets. Unfortunately, this information needs to be created and updated parallel to system development. So, if documentation is available, chances are that it is outdated and incomplete.
- The program *source code* is the most accurate and up-to-date information about a software system. Unfortunately, the mental reconstruction of abstractions that are expressed via programming language constructs is very challenging. Any program control flow and its effects can be distributed in space and time, which poses high expectations on developers' imagination abilities as well as working memory and short-term memory.
- The *running program* itself enables developers to experience concrete behavior by making fewer demands on mental capabilities. Unfortunately, the connection between observable run-time behavior and corresponding source code can be challenging.

Developers need access to information about the concrete run-time behavior of a program. As abstract descriptions require interpretation, they would like to reduce error-prone guesswork and access as much concrete data as possible. They would like to work with specific objects, situations, and scenarios like the ones the authors of the particular piece of code had in mind. Although developers ease this comprehension activity by drawing on their former experiences, they keep on looking for clues to identify program behavior that deviates from the default case. By doing so, many questions arise, which could be answered with the help of run-time information (Sillito et al. 2008), such as *How does this [...] [object] look at run-time?*

As developers continue to get a software system to know, they would like to make notes of their findings and insights to recall them later on (Fig. 2). Without that, information would have to be organized only in one's mind, and this would make it more difficult to reach new thoughts. Here externalizing existing knowledge helps. In case of interruptions, developers could use these notes to continue quickly from where they left off. Besides interruption recovery, this information offloading also reduces the mental workload and eases reflection. Once visible, the notes are intended to be freely arranged to work with appropriate points of view by means of grouping and sorting.

As developers have their own notion of an ideal comprehension process, programming environments should support this flexibility without trying to enforce a different one due to technical limitations or historical reasons. Thus, tools should provide access to concrete run-time information and enable developers to freely arrange all externalized findings. We will now illustrate these limitations of traditional programming environments.

Fig. 2 Similar to design thinking methods, developers would like to externalize their thoughts (here: using a whiteboard) to better reflect and talk about their understanding of a system part. All artifacts (here: documents, photographs, and drawings) are visible and freely arrangeable in order to reach new insights more easily

3.2   The Lack of Integrated Tool Support

Developers prefer to use integrated programming environments instead of many stand-alone tools. Although there are developers who feel comfortable using separate text editors, compilers, debuggers, and version control tools, Sillito et al. (2008) conclude that iterative and connected program comprehension questions benefit from co-operative tools combined within one environment. Therefore, we focus on arguing limitations in integrated programming environments only.

Integrated programming environments provide access to run-time information by means of program execution within a debugging mode. After entering this mode, execution stops at a selected location in the source code—a so-called breakpoint. Such a halt in the execution represents one point in time where developers can explore program state and access concrete data to better understand abstractions. Unfortunately, this level of program comprehension support is limited:

- Run-time information is not directly accessible. Developers make educated guesses to determine breakpoints, but they cannot always be sure that those are actually encountered during program execution. This trial-and-error method can and often does lead to loss of focus and increased comprehension efforts.
- There is only one single point in execution time accessible. For example, a comparison between program states at multiple points in time is difficult and requires increased mental efforts or additional tools.
- Access to context information for one point in execution time is limited. Developers can only see the direct path of method calls that led to the breakpoint. For example, there are no side effects visible that are caused by conditional branching. Technically speaking, there are no call trees, but only call stacks.

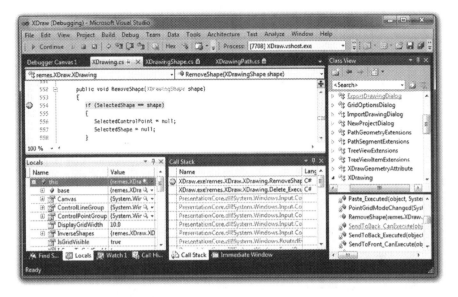

Fig. 3 Traditional programming environments (here: Visual Studio 2010) do not allow for externalizing thoughts because of the inflexible, arbitrarily tiled user interface design. In addition, they provide only limited access to run-time data by means of breakpoint debugging

Integrated programming environments need to present data in an interactive way so that developers can process it. Due to limited screen space, it is not possible to show all available static and dynamic information simultaneously. Basically, the user interface of these environments supports only document-centered workflows within the desktop metaphor. By doing so, informational units can be arranged freely like document papers or stacked conveniently like an address card index—all on a two-dimensional area mimicking a physical desk. Unfortunately, this approach dictates a comprehension process that is different than what developers would like to follow:

- Developers cannot treat all available information equally. The tiled arrangement of information is centered on a source code area (Fig. 3). Having this, developers are guided to repeatedly return to the source code view, whether or not they really need to.
- Developers cannot explicitly see relationships between visible information. All data boxes are meant to be used in an isolated fashion. Developers have to find clues by means of text labels in order to reveal connections, for example, between a call stack entry and the source code of the corresponding method executed.
- As a result, developers cannot externalize their thoughts in programming environments and interruptions could become problematic (Ko et al. 2006).

Traditional programming environments dictate a comprehension process that impedes efficient software maintenance. On the one hand, developers cannot access

Fig. 4 Automated access to run-time information: Test coverage (*left*) is used to find relevant entry points for program execution (*middle*). Each node in the accessible call tree is connected to the more detailed system state (*right*)

Fig. 5 Layout of the new concept: All kinds of information are displayed consistently in columns, for example, call trees (*left*), source code (*middle*), and object states (*right*). Overlay annotations make collaboration and dependency information explicit

valuable run-time information directly. On the other hand, they cannot process available information as conveniently as needed; graphical user interfaces are limited due to an inappropriate user interface metaphor. Externalization of thoughts is not possible and hence limited mental capacity means the evaporation of transient insights. In the end, this comprehension process promotes error-prone guesswork.

3.3 Applying Design Methods to Programming Environments

We propose *vivide*: our concept and prototypical implementation for next-generation programming environments to better support program comprehension activities according to developers' needs. This concept addresses two major issues of traditional programming environments: explicit run-time data access using a debugging mode and inappropriate visualization of accessible data.

We avoid the need for an explicit debugging mode by making use of tests to access run-time information implicitly. As complex software systems benefit from automated tests to ensure functionality (Beck 2003), we assume that the system under development is covered by tests to a certain extent. Having this, we can select all tests that cover a piece of code to access valuable run-time information when needed (Perscheid et al. 2010) (Fig. 4). In a first step, full call trees of involved test runs are captured and provide context information to developers. In a second step, calls in the tree are connected to more detailed system state. Having this, developers benefit from the fact that static source code information is directly connected to

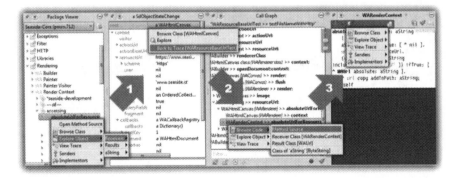

Fig. 6 Screenshot of our prototypical implementation called *vivide*. Each column offers context menus to access more static or dynamic data. With this, direct navigation between source code and run-time is supported. For example, accessing an example object of a class (*1*), browsing the run-time context (*2*), and going back to the source code (*3*)

available run-time data and vice versa. This mechanism even allows for accessing multiple points in execution time simultaneously—from a developer's perspective.

We propose a new user interface metaphor to create a visual space that allows developers to follow a better comprehension process. All information can be arranged column-wise on an unlimited horizontal space (Fig. 5). Each column displays either static source code or dynamic run-time information. Columns appear and disappear as expected while developers explore system behavior. The clear separation of horizontal and vertical screen axis supports developers' orientation: The horizontal axis shows different kinds of artifacts and the vertical one offers details for each kind.

Our new concept for integrated programming environments enables developers to experience a story—the story of a software system. The column-oriented display for static and dynamic information is supplemented with interactive navigation facilities that allow for switching directly between source code and run-time. This means, that valuable run-time information is omnipresent. Additionally, the principle of *direct manipulation* (Shneiderman and Plaisant 2009) is pursued by means of objectifying visible information. This allows developers to be more immersed in the program comprehension process because they no longer work with abstract containers that present information, but manipulate the information itself directly. They are able to externalize their mental model to better reflect on their thoughts. In the end, developers can collect information conveniently to gain the relevant knowledge that is needed to accomplish programming tasks successfully—even when facing large and complex programs (Fig. 6).

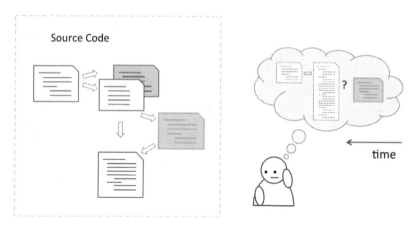

Fig. 7 While being in the middle of a refactoring, a detail becomes apparent that impacts the developers' perception of a refactoring

4 Prototyping Support for Program Design Activities

While in the middle of a change, developers are likely to get a better understanding of the existing code base and of the improvements planned. This means that the understanding of the problem co-evolves with the implementation of its solution, similar to other design fields (Dorst and Cross 2001).

As a result, developers will regularly face one of the following situations: They want to withdraw their recent changes or they want to study a previous state. In the former case, a promising idea might turn out to be unattractive or even wrong. Developers thus want to cancel recent changes and start over from a previous state of development. In the latter case, developers want to inspect and study a previous state of the code to recover some knowledge, which is currently inaccessible but important for the task at hand. Figure 7 illustrates this aspect. Developers discover an additional detail that was missing from their understanding but impacts their current idea of the refactoring.

4.1 The Lack of Tool Support

While developers often encounter the need for going back, current tools lack appropriate support to do so. Development environments and text editors in general typically provide an undo/redo feature. This feature allows developers to cancel most recent changes. It makes typing on a computer more convenient compared to using a typewriter or writing by hand as, for example, typos can be undone without requiring a complete rewrite of the current page. Such a feature works on the level of characters, as it allows undoing and redoing the insertion of individual characters. Furthermore, it handles files independently of each other, as it maintains a separate history for each of them.

While an undo/redo feature is very convenient for changing recently entered text, going back to a less recent version of a file is rather tedious. Moreover, refactoring code or implementing a feature typically requires the manipulation of various files, which cannot be undone easily as this requires developers to manually apply the undo feature repeatedly for each changed file.

Version control systems (VCS) (Spinellis 2005) can manage multiple files of a project and allow for reverting all files to a state where they were at a given point in time. During development, developers can snapshot the current state of all files that belong to the project, and the VCS will assign a version number to this snapshot. Such a snapshot, which mainly describes the difference to the previous version, can also be shared with co-workers. Having created a series of snapshots during development, developers can always go back to a previous snapshot and continue working from there.

While such a VCS allows for switching the state of a project, creating a snapshot requires conscious effort. Developers always have to consider whether they should snapshot the current state of development, that is, whether there might be a need of going back to exactly this state. Of course they don't know if this will be necessary, which renders these considerations an assessment of risks. Moreover, when one is in a creative process one typically does not think of something process-related at the same time. So, developers might miss important opportunities for taking a snapshot. And, if their subconscious reminds them to consider taking a snapshot, this is likely to interrupt creative thought and will require a significant amount of time to continue the previous task.

Furthermore, a VCS lacks direct support for reverting multiple interdependent projects. An application typically consists of multiple components that are managed as different projects, for example, to support reuse of components, which might be developed independently. In such a scenario a VCS does not help to revert the whole system as its scope is limited to one project.

Because developers don't know what snapshots might be required later, they tend to choose one of the following two approaches. Either they snapshot very often or try to avoid mistakes that would require them to go back. Both options have consequences that render them suboptimal. The former, taking snapshots continuously, has the benefit that there will always be a snapshot to go back to. However, a VCS does not make it convenient to find a particular snapshot since the only information about a snapshot is the time at which the snapshot was taken and a comment that developers have to write. Also snapshotting very often requires commenting snapshots, which is tedious.

The latter, avoiding mistakes, requires developers to contemplate ideas in detail before implementing them. They avoid making changes that are likely to not produce any meaningful result, even if they cannot be sure about it. As a result, they might decide to implement the idea or to throw the idea away. If they decide to change the code base they might forget to snapshot, and in both cases their decisions might be wrong. So, developers consider whether an idea is worthwhile

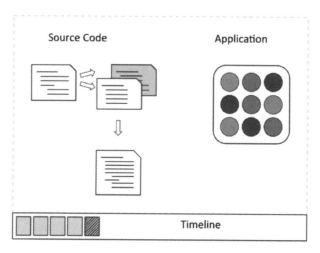

Fig. 8 The timeline metaphor visualizes a program's history

to be implemented. Also, if they decide to implement the idea, they tend to study the current state of the code base in detail to be sure they have understood all relevant aspects to not break anything.

We can observe that the lack of dedicated tool support for going back results in habits that try to avoid undesired situations but are likely to impede creative work. Whatever developers are doing, they have to continuously consider available options and their consequences: doing a snapshot or going ahead without it.

4.2 Proposed Approach

We propose *continuous versioning* to encourage a trial-and-error approach during program design activities. Continuous versioning keeps track of intermediate development states of the program, in particular those between explicit snapshots, and allows for continuing development from any of those intermediate states. It arguably encourages developers to make mistakes and to learn from them, as they can get back to a desirable state of the program with only a small effort.

The concept of continuous versioning means that the development environment takes a snapshot of the program after each modification without manual interven-tion or initiative of developers. More specifically, when developers save recent modifications the development environment will create a new version for the current state of development. Implementing a feature or trying to improve a program's design will thus create several versions. Similar to undo/redo features of traditional editors, the version history is maintained by the tools in the back-ground, without being noticed by developers. A program's history can be visualized using a timeline metaphor as shown in Fig. 8. The timeline lists all previously created versions of the program under development. Developers can undo their changes by going back to a previous version. Immediately after switching to another version, all integrated tools are updated to the content of the newly selected version.

Our proposed approach makes it easier to edit interrelated program artifacts, like the regular undo/redo feature does for editing files. Since the invention of undo/redo, the effort to correct typos has been considerably reduced and mistakes can now be corrected more easily. Similarly, continuous versioning increases the ease of writing and designing programs: When a promising idea becomes unsuitable after a couple of modifications, developers can get back to a more desirable state of the source code with only little effort.

5 Summary

In this chapter, we have described how developers are continuously involved in program redesign activities. We have illustrated some aspects of program design and explained their meaning for development tasks. We have then reported on our ongoing work of adopting design practices for programming in two respects: First, we proposed a new concept and a prototypical implementation for programming environments that should improve the program comprehension process by means of directly accessing a program's run-time information and freely arranging information on an unlimited horizontal space. Second, we have presented the concept of continuous versioning that keeps track of intermediate development states. This enables developers to continue development from any of such intermediate states, making it easier to withdraw recent changes. We can conclude that principles and practice of *Design Thinking* are of significant interest for software development in general and program design activities in particular.

References

Ashby FG, Isen AM, Turken AU (1999) A neuropsychological theory of positive affect and its influence on cognition. Psychol Rev 106(3):529

Beck K (1997) Smalltalk best practice patterns. Prentice-Hall, Upper Saddle River

Beck K (2000) Extreme programming explained: embrace change. Addison Wesley, Reading

Beck K (2003) Test-driven development: by example. Addison-Wesley Professional, Boston

Dorst K, Cross N (2001) Creativity in the design process: co-evolution of problem-solution. Des Stud 22(5):425–437

Fowler M (1999) Refactoring: improving the design of existing code. Addison-Wesley Professional, Boston

Fowler M (2011) Opportunistic refactoring. November 2011. http://martinfowler.com/bliki/OpportunisticRefactoring.html

Fowler M, Highsmith J (2001) The agile manifesto. Softw Dev Mag 9(8):29–30

Johnson RE, Opdyke WF (1993) Refactoring and aggregation. In: Object technologies for advanced software. First JSSST international symposium. Lecture notes in computer science, vol 742. Springer, November 1993, pp 264–278

Kay A et al (2010) STEPS toward expressive programming systems, 2010 progress report submitted to the national science foundation. Technical report, Viewpoints Research Institute

Ko AJ, Myers BA, Coblenz MJ, Aung HH (2006) An exploratory study of how developers seek, relate, and collect relevant information during software maintenance tasks. IEEE Trans Softw Eng 32:971–987

Lim Y-K, Stolterman E, Tenenberg J (2008) The anatomy of prototypes: prototypes as filters, prototypes as manifestations of design ideas. ACM Trans Comput Hum Interact (TOCHI) 15 (2):1–27

Meyer B (2000) Object-oriented software construction. Prentice Hall, Upper Saddle River

Opdyke WF (1992) Refactoring object-oriented frameworks. Ph.D. thesis, University of Illinois

Perscheid M, Steinert B, Hirschfeld R, Geller F, Haupt M (2010) Immediacy through interactivity: online analysis of run-time behavior. In: 17th working conference on reverse engineering. IEEE, Beverly, pp 77–86

Ralph P, Wand Y (2009) A proposal for a formal definition of the design concept. In: Design requirements engineering: a ten-year perspective. Springer, Berlin/Heidelberg, pp 103–136

Schwaber K, Beedle M (2001) Agile software development with scrum, 1st edn. Alan R. Apt

Shneiderman B, Plaisant C (2009) Designing the user interface: strategies for effective human-computer interaction, 5th edn. Addison Wesley, Reading

Sillito J, Murphy GC, De Volder K (2008) Asking and answering questions during a programming change task. IEEE Trans Softw Eng 34:434–451

Simon HA (1996) The sciences of the artificial. The MIT Press, Cambridge, MA

Spinellis D (2005) Version control systems. IEEE Softw 22:108–109

Wikipedia: the free encyclopedia (2011) Design. November 2011, http://en.wikipedia.org/wiki/Design

Virtual Multi-User Software Prototypes III

Gregor Gabrysiak, Holger Giese, and Thomas Beyhl

Abstract In design thinking and software engineering, prototypes play a crucial role in validating insights, needs and requirements. Still, the effort necessary to create these prototypes depends on multiple factors such as the number of people involved with the design thinking project. Especially for multi-user software systems, the effort of creating validation artifacts is too high to be feasible for multiple iterations, thus, inhibiting design thinkers to inexpensively try different ideas early and often. To overcome this problem, we investigated the usability and feasibility of virtual prototypes – animated simulations of formal models which can be derived automatically without additional costs. This paper reports on our advances during the 3 years of the design thinking research project concerned with Virtual Multi-User Software Prototypes.

1 Introduction

Design thinking is characterized as a highly interactive and incremental process. People with different backgrounds and experiences are crucial to drive it. This heterogeneity is important to include as many viewpoints as possible during the need finding and the design, not only the perceptions of technical experts and engineers. Due to these diverse backgrounds, prototypes are essential in establishing a common understanding between all persons involved or impacted by a design thinking project.

In software engineering and engineering, models play a key role in designing complex systems. A model is an artifact that abstracts from all aspects which are considered irrelevant for a specific purpose, thus, allowing observations of relevant

G. Gabrysiak • H. Giese (✉) • T. Beyhl
System Analysis and Modeling Group, Hasso Plattner Institute for IT Systems Engineering at the University of Potsdam, Prof.-Dr.-Helmert-Street 2-3, D-14482 Potsdam, Germany
e-mail: holger.giese@hpi.uni-potsdam.de

H. Plattner et al. (eds.), *Design Thinking Research*, Understanding Innovation, DOI 10.1007/978-3-642-31991-4_15, © Springer-Verlag Berlin Heidelberg 2012

aspects only (Stachowiak 1973). But modeling is not restricted to engineering; also prototypes in a design process are models (Gabrysiak et al. 2010a).

In many engineering disciplines, simulation and prototyping are employed on a regular basis to better understand the problem at hand and identify innovative solutions. When designing, e.g., a graphical user interface (GUI) of a software system, paper prototypes can be used, so-called low fidelity prototypes (Snyder 2003). Once the user is satisfied with the low fidelity prototype, a high fidelity prototype is developed to enable the discussion of more detailed issues. However, today the same is not possible for complex multi-user systems and their overall behavior since they are not only dependent on how individual end users interact with the respective GUI, but mainly on how the users interact with each other.

Still, to create a useful innovative product, it is crucial to understand the real needs of the end users. Therefore, the design thinkers have to gather insights about the end users and their domain. More specifically, what do end users do, how do they do it and why is it done. After design thinkers have gathered assumptions about the end users, it is imperative to validate what was captured as well as the subsequent design ideas. Producing tangible prototypes has proven to be an essential tool for establishing a common understanding within heterogeneous design teams, but also with and among end users, about what the design thinkers learned. By using prototypes to elicit or provoke feedback about assumptions gathered beforehand or manifestations of design ideas (Lim et al. 2008), it becomes possible to validate the assumptions and the resulting design choices.

Tangible prototypes support the design thinkers' need to gather feedback from end users to validate the underlying concepts they embody. In typical design thinking projects these prototypes can be derived straightforwardly which allows validating concepts quite early, inexpensively, and often. This allows design thinkers to *fail early and fail often* instead of spending too much effort and resources on the creation of expensive validation artifacts.

As pointed out by Andriole (1994), prototyping is very effective also for software systems, since it allows end users to experience something tangible and provide feedback about it. For this domain, Bäumer et al. identify three different kinds of prototypes for software systems (Bäumer et al. 1996). While *Explorative Prototypes* are commonly used at the beginning of a project to test people's reactions to new concepts, *Experimental Prototypes* are produced to evaluate whether a concept fulfills the end users' expectations. *Evolutionary Prototypes*, on the other hand, combine both approaches. Through multiple iterations, the prototype matures till the final prototype can be considered as the final result.

Typical design thinking software projects consequently employ prototypes of graphical user interfaces (GUI) where end users are in the center of attention (Winograd 1996). Different approaches of producing representations of the captured needs and developed concepts suitable for end users are illustrated in Fig. 1. However, for complex multi-user software systems this is not the case. Since GUI prototypes usually only represent a single user's view on the system, their applicability is limited to validating the design of this individual view (*how can I do what*). Thus, the validation of the underlying rationale (*why do I do it*) is not feasible for all

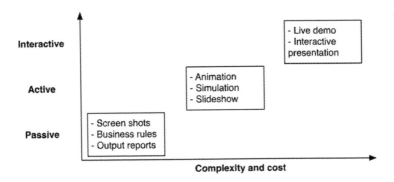

Fig. 1 Storyboarding and prototyping software requirements (Adapted from Leffingwell and Widrig (1999))

involved end users. Looking at the complete system, however, building a prototype that captures the whole underlying processes and data is prohibitively expensive. The high costs associated with the production of such prototypes of complex software systems imply that only few prototypes would be economically feasible, if any at all. Consequently, in software engineering for complex multi-user software systems, end users' needs and potential solutions can only be described at a high level of abstraction using software engineering models such as processes and scenarios, but not with testable prototypes. Still, design thinking strongly depends on insights about the design challenge gathered directly from end users interacting with prototypes. Prototyping allows obtaining further insight due to end users' feedback and to end up with a suitable solution for the right problem, if possible chosen from multiple design ideas (Tohidi et al. 2006). Problems arise especially when design thinkers have to design solutions involving multiple users as each end user's understanding of the problem and the implicated individual needs may not only be different but even conflicting. Therefore, besides eliciting insights from all end users, it is also mandatory to resolve these conflicts by creating a common understanding of the problem domain, which can be quite complicated and costly.

In addition, not all members of the design team might be familiar with the employed software engineering models and, thus, can hardly contribute to them in a direct manner. Up until now, multiple iterations with tangible prototypes are not feasible for complex multi-user software systems. Thus, design thinking cannot realize its full potential when addressing such problems.

Together with our industrial partner D-LABS GmbH, a design consultancy for software products, we investigated how they adapted the design thinking process to deal with the intangibility of software systems. Based on these insights, we came up with an approach of combining the formality which is usually required by software engineers to deal with inconsistencies with the ability to quickly validate what has already been understood by means of prototyping. To be able to fail early and often, the costs associated with such a prototype have to be minimal. This is critical in a domain in which prototyping is associated either with graphical user interfaces, which can simply be sketched, or evolutionary prototypes, which are too expensive to throw away. Instead, our approach of virtual prototypes exploits the formality of behavioral specifications within the domain of multi-user processes through enactment and visual metaphors that end users are familiar with.

In this chapter, our approach for prototyping multi-user software processes in a scenario-based and tangible manner is presented. In Sect. 2 we describe related approaches. Then, Sect. 3 discusses our approach while Sect. 4 describes the ongoing evaluation. Related research activities within the context of this project are outlined in Sect. 5. Finally, we summarize our results and provide an outline for future work in Sect. 6.

2 State of the Art

Many different modeling approaches exist to model the end user requirements for complex software systems, each restricted to capture only specific aspects (Alexander 2011). However, most of these models are not intuitive enough to empower end users to judge and comment on the correctness of the captured insights.

Depending on the type of software system, the corresponding requirements are potentially hard to communicate. Requirements of software systems which are directly used by end users are quite suitable to be prototyped quickly and inexpensively. However, in domains such as safety-critical systems the requirements are either too abstract or too complex to be prototyped in such a fashion. The only feasible kind of prototype for this category of systems needs to be (semi-) automatically derived (Pohl 2010). Otherwise, the costs of the creation might exceed its value during the validation. Another dimension which influences the feasibility of prototyping is the size and duration of the project. In small-scale projects, agile approaches are centered around evolutionary prototypes, which are not only used for validation but which are also incrementally refined till they are accepted as the final product.

However, as Alexander (Alexander and Beck 2007) put it, "It can't always be right to develop the requirements during coding – still less, while concrete is being poured and metal is being cut. But it's certainly right to select a life cycle that encourages dialogue, with iteration in *baby steps*." So, depending on the dimension of a project and how many end users are in the center of the project, different approaches are more suitable than others. For software intensive systems in multi-user environments, model-based approaches aim to iteratively refine a model which is then used to derive other design artifacts. The formal models used in these domains are not understandable for end users. Consequently, some approaches have been proposed to ease these communication problems. Apart from being explained to everybody in each validation, models can either be annotated to be more self-explanatory (Al-Rawas and Easterbrook 1996) or different states can be visualized (Özcan et al. 1998). Additionally, a combination of both options is possible (Haumer et al. 1998). However, all of these approaches require the modeler to create custom-made additions to the models. For this case, Pohl points out that prototyping is only feasible if the prototype can be (semi-) automatically derived from the underlying models (Pohl 2010). Storyboarding (Sutherland and Maiden 2010) is quite suitable to validate insights early as well as later during a design thinking project. However, the manual creation of such a prototype usually requires too much effort to be feasible for complex systems.

Harel and Marelly's *Play*-Engine (Harel and Marelly 2003) allows end users to specify the behavior of reactive software systems in terms of scenarios. These scenarios can be simulated, i.e. played-out, so that end users are able to validate the correctness of the synthesized sequences without modeling knowledge. Their approach allows end users to directly play-in additional scenarios by interacting with an intuitive user interface. Still, this approach is restricted to capture interactions between the end user and the software that is designed. Due to the focus on system reactions on user actions implemented in the play engine, interactions between the system and several users or among users cannot be simulated yet. Consequently, it is not suitable for complex multi-user workflows.

3   Approach

Our approach tries to bridge the gap between the formal requirements models used by practitioners and end users via prototypes. As sketched in Fig. 1, such an approach is only successful, if the prototypes can be derived directly from the formal representation to be feasible (Pohl 2010). At the same time, they need to be intuitively understandable for the end users (Leffingwell and Widrig 1999).

Figure 2 illustrates a sequence of activities, which represents our methodology of validating insights and needs of end users via virtual prototyping of software engineering models. Initially, end users are interviewed by design thinkers to gather insights and needs (1). These insights are subsequently externalized into software engineering models by design thinkers (2). Formal models are necessary to cope with the inherent complexity of multi-user software systems. While word documents can be used to collect all requirements, they do not scale up to be feasible to maintain multiple thousands of requirements distributed over several hundred pages. To validate insights on how end users work and their needs, our approach automatically derives a virtual prototype based on these models as a tangible representation suitable for the end users (3). These virtual prototypes are closer to the domain of expertise of end users and, thus, can be experienced by end users directly (4). Since not all Design Thinkers are familiar with models specifying insights about software systems, they can also experience the virtual prototypes just as the end users. While experiencing the virtual prototype, end users perceive concepts embodied in the virtual prototype and compare them to their corresponding perception of their domain (5) (cf. (Pohl 2010)). This allows them to appropriately provide feedback about concepts within their domain of expertise (6). Apart from this feedback, end users directly validate all models that affect them – if they (implicitly) agree, the

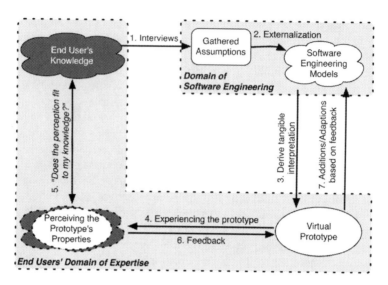

Fig. 2 How virtual prototypes are created and how they allow end users to comment on the models

models that were presented to them are considered as validated. Otherwise, they can play-in the correct behavior themselves as part of their interaction with the interactive virtual prototype. Additionally, more complex changes still need to be realized through a modeler (7). This sequence can be systematically repeated until a common understanding of the problem domain among all persons involved in the design thinking project is achieved.

The individual parts of the prototypical implementation of our approach, the simulator and the animation which is suitable for the general domain of multi-user processes and how they interlink are discussed in (Gabrysiak 2011). While the simulator progresses through a process, the animation is updated accordingly.

3.1   Capturing Insights in Software Engineering Models

While software engineering focuses on how to correctly build the software, requirements, engineering ensures that the right software is built. To achieve this, communication is essential to describe what end users require, since the ones eliciting their needs are usually not the ones who end up engineering the solution. Consequently, all gathered insights need to be communicated as completely, correctly and consistently as possible, since the engineers realizing the software do not have the same insights, or even experiences as the persons who collected this knowledge.

A multitude of approaches for capturing the needs of end users exists (Fig. 3), each one with its particular advantages and shortcomings (Alexander 2011).

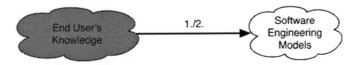

Fig. 3 Knowledge about who does what how needs to be captured systematically

Still, the possibility of validating models created with these techniques is an open issue under investigation. For prototypes to be successful, they need to offer affordances that enable end users to correctly perceive the intention of the modeler without being hindered by unimportant aspects of the prototype. As we stated in (Gabrysiak et al. 2010a), models are prototypes offering affordances that only modeling experts can recognize and work with. They are capable of adding elements or removing them, just as any domain expert would be able to adjust a tangible prototype of concepts related to the corresponding domain. The modelers, however, are not able to refine or correct wrong assumptions from their models or prototypes themselves, since they do not have the necessary domain experience – this can only be achieved by the end users.

As argued by van der Aalst (2007), people tend to simplify the complexity of models they create. Unfortunately, this hides away the reality and reduces complexity for the sake of understandability. A realistic model as derived from logging information can be seen in Fig. 4. Such a model is too complex and usually combines too many different perspectives to be understandable for potential end users of a system supporting their workflows. To create an understandable representation of the gathered knowledge, this model would have to be broken down for each individual (group of) end user(s): admission for the nurses, preparation for the surgeons, and follow-up procedures for other doctors. Only by allowing all stakeholders to experience the specific parts of the multi-user process that they are familiar with, the validation becomes directed and, thus, effective.

Findings. By now, practitioners engaged in early, required engineering stages acknowledge the need for more flexible tools which offer flexibility where other tools restrict modeling efforts (Ossher et al. 2010). Usually, one is faced with the decision of either using simple, informal models or complex formal models. The more formal these models are the more support they provide to the modelers. To allow the creation of a feasible virtual prototype, the employed models need to be executable.

Also, the models need to contain side-effects which can be mapped directly into the domain of expertise of the end users. Only then, a visualization can be found which is intuitive enough to convey the modeled intention to the end users.

Results. Apart from coming up with a flexible way of modeling the necessary information of who does what and why (Gabrysiak et al. 2010d), we also categorized different ways of dealing with this trade-off between formal tool support and being able to express the gathered knowledge, i.e. potentially *everything*, during the modeling stages (Gabrysiak et al. 2011a).

Fig. 4 A process model
based on log files representing
the flow of 874 patients
having heart surgery
(van der Aalst 2007)

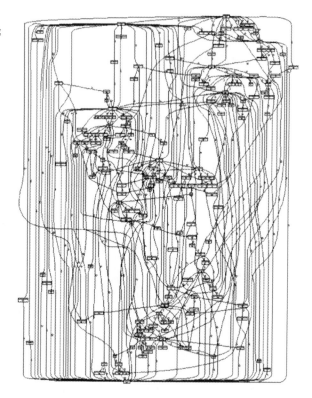

Our approach relies on a graph transformation system as proposed by Rensink (2010). By specifying the insights as graph transformations, they are easily modifiable and, thus, flexible. Also, they are specifically designed to describe structural changes via patterns. This allows the modeler to capture just how the context an end user perceives changes based on actions or interactions with other end users.

The prototype we implemented provides most of the required features. The balance between formality and understandability was decided in favor of formal models. Thus, the formal models which enable us to derive the virtual prototype are not intuitively understandable by end users or even design thinkers engaged in collecting the insights. However, a master's thesis concerned with the creation of a modifiable natural language representation for these models is currently in progress.

3.2 Virtual Prototypes

Findings. For virtual prototypes to be effective, they need to be feasible. They also have to provide end users an intuitive interactive representation of what was modeled, thereby allowing them to perceive and evaluate the virtual prototype

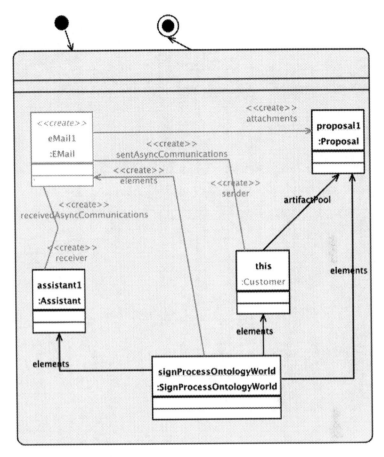

Fig. 5 Formal model capable of realistically capturing the context and the rationale behind decisions of an end user

(5. in Fig. 6). This promotes the possibility to explore other aspects of what the virtual prototype embodies.

After examining which kinds of information D-LABS recorded and relied on during their design thinking software projects (Gabrysiak et al. 2010c), we investigated a feasible solution to the problem of how to validate the complex models necessary to realistically represent *what* is *why* done by *whom*. As pointed out already, the feasibility of prototyping depends on the costs of the creation of each prototype. Thus, to be able to automate most of the steps during the creation of a prototype, we chose a formal modeling approach (an example is presented in Fig. 5). Such a prototype is not tangible in the same sense as a clay model of a tea cup. Instead, it enables end users to experience interactions with other (potentially simulated) end users just as they experience it on a daily basis. This allows them to point out inconsistencies between their daily experience and what is presented to them during their interaction with the virtual prototype. Thus, they can experience

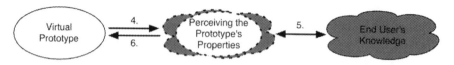

Fig. 6 A virtual prototype needs to be perceived intuitively to compare it against someone's own perception of reality

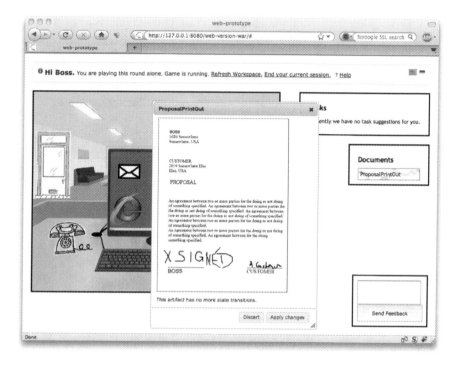

Fig. 7 Screenshot of a situation in which the participant of a simulation interacts with a document

how they were understood by the modeler and evaluate whether the modelers got it right.

The best way of presenting the understanding of what is done by whom is to be able to directly ask *why* it is done. During a validation, this can be achieved with an interactive representation instead of a passive explanation of a model (cf. Fig. 7). An interactive prototype of what was understood allows end users not only to perceive how they were understood, but also to provide new insights based on their interaction with the prototype (Fig. 6). Similar to how an end user presents an unexpected use case for a physical prototype, interactive prototypes should offer similar degrees of freedom for the end users. We refer to a prototype which offers these capabilities as a virtual prototype (Gabrysiak et al. 2010a). The usability and thus feasibility of a virtual prototype highly depends on whether it is intuitive enough to be used by end users (Gabrysiak et al. 2009).

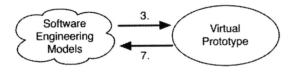

Fig. 8 The virtual prototype is derived from the models and interactions with the prototype can be played back into the models

Table 1 Results of three associated master's theses

Smart simulation (Teusner 2011)	Optimized strategies	Monitoring
Strategy-based simulation (Richter 2011)	Strategy-driven execution	Preview computation
Efficient simulation (Kleff 2011)	Process navigation	Efficient execution

3.3 From Models to a Virtual Prototype and Back

Findings. For the prototypes to be feasible, they should be automatically derived from the models. Additionally, by directly interpreting the models, no indirection is introduced during the execution – everything happens just as specified in the model. Furthermore, the new models should be added intuitively so that end users are capable of doing so by themselves, e.g., through the interaction with a derived virtual prototype (*3.* in Fig. 8) by playing in (*7.* in Fig. 8).

To achieve an interactive presentation of a model, it needs to be executable. Only then we are able to simulate the behavior previously observed from other end users who are either co-workers or other participants in the process under investigation. Having such behavioral specifications, the simulator is capable of responding to each already known and, thus, expected stimuli from the end users by creating an output for the end users participating in the simulation. Each reaction of the simulation is a piece of the process puzzle that the participating end users either agree or disagree on. In the former case, the model specifying the behavior of another end user that was just executed can be considered correct and validated from the perspective of the participant of the simulation. If that person disagrees, however, the corresponding model is considered incorrect and needs to be fixed to fulfill the participant's expectations, i.e. how they perceive reality. The main benefit of our approach is the fact, that the end users directly interact with and can comment on the models involved during the simulation (Gabrysiak et al. 2010b). The following paragraphs are based on the master theses listed in Table 1.

Monitoring. During the execution of a virtual prototype, some states should be avoided due to undesirable conditions or violations of predefined conditions. On the other hand, other states need to be reached for a successful execution of the simulation, e.g., that a customer must receive an ordered item. A monitoring approach capable of restricting the execution of the virtual prototype was included. As illustrated in Fig. 9, for an order to be successfully shipped to a customer not more than two time units, i.e. `ticks`, may pass. Consequently, the second state has to be avoided during a simulation aimed at successfully completing an order.

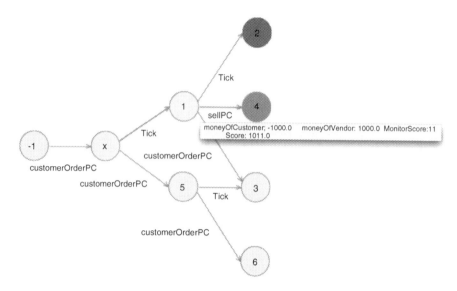

Fig. 9 In this process of selling a computer, the optimal path (`Tick` and `sellPC`) is chosen based on calculated scores which are presented as a tooltip (Teusner 2011)

Additionally, resource consumption was included into the strategy-based simulation to enable the modeler or the participants to find better ways of going through the multi-user workflow under investigation (Teusner 2011). In Fig. 9, the most successful alternative based on the monitored resources is the fourth state which is the only one where the transaction is completed. Another, more sophisticated example is illustrated in Fig. 11; the process modeled therein is predicted for 12 steps up-front. The corresponding interface provided for the modeler is presented in Fig. 10.

Consistency. The models might include inconsistencies, especially as gathering new insights might invalidate older ones or make them obsolete (Pohl 1993). These inconsistencies can be identified during a simulation or by automatically checking the models, whether specific situations can be achieved (a successful completion of all goals) or avoided (failing to achieve these goals). Consequently, to increase the chance of finding all errors and inconsistencies, a strategy-based execution mode for the simulation of virtual prototypes was implemented (Richter 2011). By choosing a strategy that tries to successfully achieve all goals for the investigated multi-user process, the simulation can already be used with an early, incomplete set of specifications to explore different scenarios (Harel et al. 2002). Later, after an allegedly complete set of specifications has been collected, another strategy can be employed to find inconsistencies, which might result in inconsistent states during the simulation.

Performance. The speed of the execution of the simulation is of the upmost importance. To be intuitive and generally usable, the underlying simulation has to be fast enough not to hinder the end users' participation. The performance of the

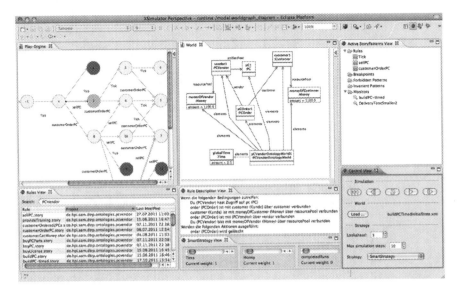

Fig. 10 The modeler's interface for steering a simulation (Teusner 2011) as opposed to the end users' perspective in Fig. 7

simulation was significantly improved as part of a master's thesis (Kleff 2011). Still, the end users have to be able to observe all relevant changes.

The resulting workflow is illustrated in Fig. 12. The current state of the simulation (*left side*) is visualized for all end users. For this state, the available behavioral specifications are tested to find suitable matches. To simulate one of the involved end user roles, the behavior of the corresponding role can be executed to arrive at the respective follow-up state (*right side*). All changes affecting the participating end users can then be visualized for them during this play-out of behavior. The play-in works the other way around – allowing end users to interact with the user interface while a specification of what was observed is created (*top*).

4   Evaluation

The approach presented in this chapter was evaluated in multiple increments. In end user studies with secretaries at the HPI, the rough idea was initially evaluated in qualitative interviews. The overall feedback already suggested that the concept would be helpful in validating insights into their workflows.

Another iteration was tested in a setting with 36 modeling students and nine potential end users. In groups of four, the students interviewed one end user, specified their findings using either (a) our approach of interacting with the end user in a virtual prototype, (b) a formal UML modeling tool or (c) an approach similar to design thinking in which the students were granted pens, paper, and

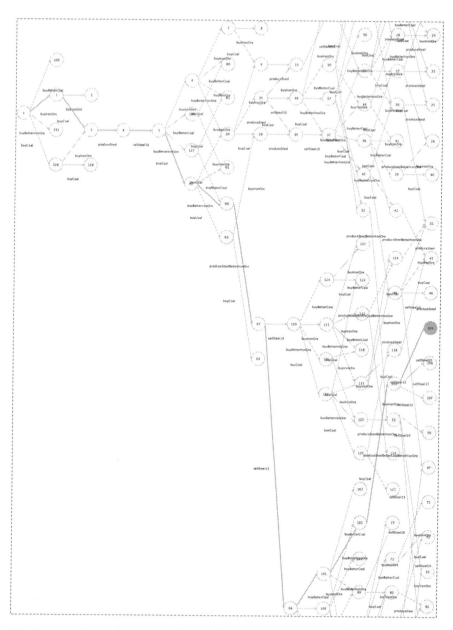

Fig. 11 An abridged preview based on more than 120 different states used to predict the best path (*thick arrows*) for a steel forge example (Teusner 2011)

complete freedom. Their results were then validated with the end user. Our results indicated, that the students and the end users were not only able to work with our approach, but the students using it also needed less time in comparison to the other groups (Gabrysiak et al. 2011b). In general, while offering all the capabilities of

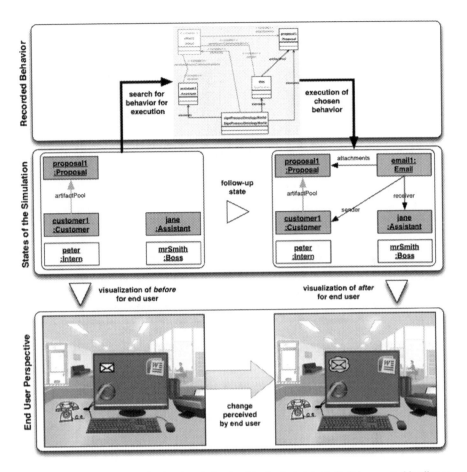

Fig. 12 By matching the behavior description (*top*) in the first state (*left*), it is executed leading to the follow-up state (*right*) which is visualized accordingly for the end user (*bottom*)

formal modeling techniques, the prototype scored better than a purely formal modeling approach and only slightly worse than an informal approach (cf. Fig. 13).

We also conducted a study to evaluate whether the chosen visual metaphors (e.g., a door to visit someone, a telephone offering an instant messaging window to represent calls) are intuitive to understanding for end users. Thus, we prepared seven videos in which someone interacted with the user interface to fulfill a distinct task, e.g., reading an email after getting notified about it or accepting a phone call and "talking" about a coffee break. Each video was shown twice to 15 students from other domains than software engineering. After each video, they were asked to describe what they saw happening as well as how sure they were of their interpretation using a seven-point Likert scale with 1 being very uncertain and 7 being very sure. Afterwards, their written interpretation was evaluated by four HPI students to ensure an objective rating of whether their description was suitable.

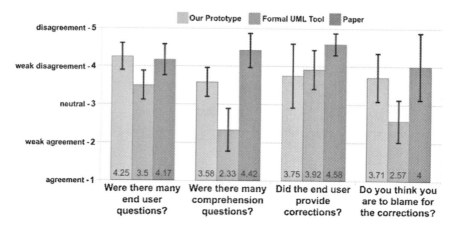

Fig. 13 Our prototype scored better than a formal modeling tool but slightly worse than the most flexible pen and paper approach (Gabrysiak et al. 2011b)

Table 2 Results for the recognition of the communication primitive using seven videos (n = 15)

	Recognition by students	Certainty *(recognized)*
1. Logging into simulation	93 % (14)	5.36 ± 1.21
2. Receiving and reading an email	87 % (13)	5.67 ± 0.67
3. Receive and answer a call	100 % (15)	6.00 ± 0.87
4. Create a document	100 % (15)	5.27 ± 1.01
5. Interacting with a document within the virtual prototype	100 % (15)	5.73 ± 1.16
6. Visiting another simulation participant	80 % (12)	6.00 ± 0.95
7. Sending an email with an attachment	100 % (15)	5.91 ± 1.22

The preliminary results are presented in Table 2. Most of the 15 subjects interpreted the interactions of an end user with the virtual prototype correctly. The worst recognition rate for a single video is 12 correct answers for the sixth video (cf. Table 2). The students providing correct interpretations were on average quite sure of their interpretation – just as sure as the students who did not understand the interactions.

Additionally, four different case studies were successfully modeled and evaluated relying on the simulator underlying our approach in three master's theses as presented in Table 1 (Kleff 2011; Richter 2011; Teusner 2011). Further evaluations are still ongoing and will be reported on later.

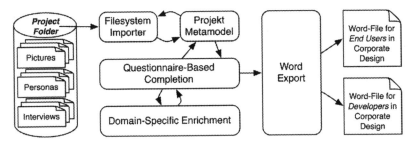

Fig. 14 The resulting tool of the bachelor project, capable of producing specifications suitable for multiple distinct perspectives

5 Additional Results

While our approach tries to solve the problem of feasible prototyping, other artifacts, which are gathered, need to be managed and organized as well. It is especially important to keep them available to derive new insights from them later on during the design thinking process. How this can be achieved was investigated in a bachelor project which aimed at combining requirements engineering with design thinking for software as it is practiced at D-LABS. This section also reports on a lecture setup which was created alongside our experiments.

5.1 Supporting Design Thinkers to Create Software Specifications

Based on our examination into how D-LABS gathers and manages their insights, a bachelor project was started to prototype a software system which combines D-LABS' Design Thinking approach for software with common requirements engineering standards. The result is illustrated in Fig. 14.

As in all projects, D-LABS uses a shared folder for all documents such as pictures, persona descriptions and digital prototypes. These folders are usually structured similarly. The students deduced a corresponding metamodel capturing most of the associations within a project, e.g., each iteration can have multiple prototype artifacts which in turn can be represented by either pictures and wireframes or executables. Thus, by looking into a project folder it is possible to infer the state of this project: the number of interviews, the number of prototypes and iterations as well as the number of presentations in-between. Of course, it then becomes possible to even point out which kind of documentation is not sufficient yet. This step can now be done semi-automatically: the prototype recognizes different file types and is able to determine the function of the documents based on their location within the folder and other properties. For instance, all JPG-files in a subfolder of "./prototypes/" usually belong to the same iteration, while the number of iterations can be derived from the number of subfolders. Considering this folder and all the documents inside, D-LABS still needs to synthesize a final presentation and handover document for all gathered information. To assist D-LABS in completing their project documentation a questionnaire-based approach is used to determine the right questions to ask (Harzmann 2011).

Additionally, depending on the kind of application that is being designed, other concerns arise that have to be dealt with. While a desktop application is usually only used by one person, an online tool needs to be suitable to serve multiple, if not thousands of users simultaneously. Consequently, the students realized a way of asking the most important questions, which strongly influence the implementation later (Knolle 2011).

From this information a final document can automatically be compiled that already contains most of the important information. This document can be generated for different stakeholders, e.g., concerning the amount of technical details needed – none for end users or all for developers. These documents are generated with the corporate identity design of D-LABS (Tietz 2011). As no approach is likely to produce a perfect document, it was important to generate a document which can be revised by somebody working on the project later on.

Alongside the bachelor project, a study was conducted with software engineering students to evaluate the acceptance of design thinking deliverables from the perspective of engineers who potentially might have to build what is specified by these deliverables. As argued by Lyon (Lyon 2011), it is important for designers to have a process that they can rely upon when explaining how they arrived at their results. In Knolle's study, most of the students did in fact not look at the description of the design thinking process which was described as the basis of the document presented to the subjects. Instead they pointed out that they trust the results of consultants if the process is specified (Knolle 2011).

5.2 Teaching

During the preparation of the evaluation, we recognized the need to recruit suitable subjects for our experimental settings. These subjects must be involved in a multi-user process which can be elicited and validated realistically. This problem is strongly connected with the problem of teaching requirements engineering to students. After an initial recruitment assessment of students who are extensively briefed to enact such a role (Gabrysiak et al. 2010e), we came to the conclusion, that at least teaching goals can realistically be achieved.

To allow students to experience the usual problems related to talking to end users, we proposed different settings suitable for such experiences (Gabrysiak et al. 2011c), one of them illustrated in Fig. 15. As a side-effect of the discussed design thinking research project, we also established a seminar format which has by now been offered twice – both times overflowing with interested students.

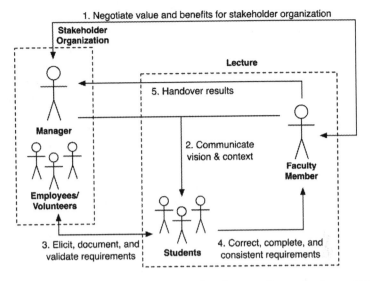

Fig. 15 To learn the need for need-finding, the students have to talk to real end users (Gabrysiak et al. 2011c)

6 Summary and Future Work

Prototypes are an important method of validating insights in design thinking projects. During the first year of the project, we investigated different possibilities of how the validation of insights could benefit from more direct end user involvement in the design of multi-user software systems. The concept we prototyped relied on executable formal models combined with a domain-specific interactive animation. Allowing end users to directly provide feedback about the model during such a simulation can be used to iteratively evolve the underlying models. This allows all persons involved in the design thinking project to develop a common understanding of what was understood by the design thinkers.

In the second year, the approach was extended with additional capabilities. Most importantly, the underlying model was replaced with a graph transformation system, capable of a more detailed specification of behavior. This also allowed the introduction of an automatic play-in mechanism along with a refinement of the existing play-out of already captured behavior. Also, our approach was evaluated within an experiment as described in (Gabrysiak et al. 2011b).

During the third year, the possibility of controlling the simulation based on the amount of decisions within a multi-user process under investigation (Richter 2011) or based on the resource consumption (Teusner 2011) was explored. Also, the overall performance during the computation of the optimal way was significantly improved as part of a third master's thesis (Kleff 2011).

By executing the models and monitoring their effects, possible conflicts can be spotted. Then, if a corresponding end user joins the simulation, the identified conflict can be resolved by guiding the simulation and the user to the according situation in which the conflict occurs. Overall, through such a guided simulation, the exploration and validation of who does what and why becomes faster and more directed, since the designers can actively set the priorities for the simulation via strategies instead of the participating end users.

The prototype we developed for our approach is suitable for software systems supporting end users in multi-user processes. For other domains, different visualizations are necessary to animate the concepts of these domains to enable their end users to understand them. Apart from the discussed experiments, the modeling approach was used and evaluated in three master's theses (Kleff 2011; Richter 2011; Teusner 2011). However, additional evaluation is still ongoing and will be reported at a later date.

Acknowledgements The authors are grateful for the input of Alexander Renneberg (D-LABS GmbH), Nico Rehwaldt, Henrik Steudel, Alexander Lüders, Stefan Kleff, Stefan Richter, Ralf Teusner, Christoph Kühnl and Jonathan A. Edelman.

References

Alexander I (2011) GORE, SORE, or what? IEEE Softw 28:8–10

Alexander I, Beck K (2007) Point/counterpoint. IEEE Softw 24:62–65

Al-Rawas A, Easterbrook S (1996) Communication problems in requirements engineering: a field study. In: Proceedings of the first Westminster conference on professional awareness in software engineering, Royal Society, London, 1–2 Feb 1996

Andriole SJ (1994) Fast, cheap requirements: prototype, or else! IEEE Softw 11(2):85–87

Bäumer D, Bischofberger WR, Lichter H, Züllighoven H (1996) User interface prototyping— concepts, tools, and experience. In: ICSE'96: Proceedings of the 18th international conference on software engineering. IEEE Computer Society, Washington, DC, pp 532–541

Gabrysiak G (2011) Exploration and validation through animation of scenario specifications. In: Doctoral symposium of the 19th IEEE international requirements engineering conference (RE'11), Trento, 29 Aug 2011

Gabrysiak G, Giese H, Seibel A (2009) Interactive visualization for elicitation and validation of requirements with scenario-based prototyping. In: Proceedings of the 4th international workshop on requirements engineering visualization, RE'09. IEEE Computer Society, Los Alamitos, pp 41–45

Gabrysiak G, Edelman JA, Giese H, Seibel A (2010a) How tangible can virtual prototypes be? In: Proceedings of the 8th design thinking research symposium, Sydney, 19–20 Oct 2010, pp 163–174

Gabrysiak G, Giese H, Seibel A (2010b) Deriving behavior of multi-user processes from interactive requirements validation. In: Proceedings of the IEEE/ACM international conference on automated software engineering, ASE'10, ACM, Antwerp, Sep 2010, pp 355–356

Gabrysiak G, Giese H, Seibel A (2010c) Towards next generation design thinking: scenario-based prototyping for designing complex software systems with multiple users. In: Plattner H, Meinel C, Leifer L (eds) Design thinking: understand – improve – apply, Understanding innovation. Springer, Berlin/Heidelberg, pp 219–236

Gabrysiak G, Giese H, Seibel A (2010d) Using ontologies for flexibly specifying multi-user processes. In: Proceedings of ICSE 2010 workshop on flexible modeling tools, Cape Town, 2 May 2010

Gabrysiak G, Giese H, Seibel A, Neumann S (2010e) Teaching requirements engineering with virtual stakeholders without software engineering knowledge. In: Proceedings of the 5th international workshop on requirements engineering education and training, RE'10, Sydney, pp 36–45

Gabrysiak G, Giese H, Lüders A, Seibel A (2011a) How can metamodels be used flexibly? In: Proceedings of ICSE 2011 workshop on flexible modeling tools, Waikiki/Honolulu, 22 May 2011

Gabrysiak G, Giese H, Seibel A (2011b) Towards next-generation design thinking II: virtual multi-user software prototypes. In: Plattner H, Meinel C, Leifer L (eds) Design thinking research: studying co-creation in practice, Understanding innovation. Springer, Berlin/Heidelberg, pp 109–128

Gabrysiak G, Giese H, Seibel A (2011c) Why should i help you to teach requirements engineering? In: Proceedings of the 6th international workshop on requirements engineering education and training, RE'11, Trento

Harel D, Marelly R (2003) Come, let's play: scenario-based programming using LSC's and the play-engine. Springer, New York/Secaucus

Harel D, Kugler H, Marelly R, Pnueli A (2002) Smart playout of behavioral requirements. In: FMCAD'02: Proceedings of the 4th international conference on formal methods in computer-aided design. Springer, London, pp 378–398

Harzmann J (2011) Fragenbasierte Vervollständigung von Modellen. Bachelor's thesis, Hasso Plattner Institute, University of Potsdam

Haumer P, Pohl K, Weidenhaupt K (1998) Requirements elicitation and validation with real world scenes. IEEE Trans Softw Eng 24(12):1036–1054

Kleff S (2011) Effiziente Simulation von virtuellen Prototypen. Master's thesis, Hasso Plattner Institute at the University of Potsdam

Knolle L (2011) Vervollständigung von Anforderungen in Software-Design-Projekten. Bachelor's thesis, Hasso Plattner Institute, University of Potsdam

Leffingwell D, Widrig D (1999) Managing software requirements. Addison Wesley, Harlow

Lim Y-K, Stolterman E, Tenenberg J (2008) The anatomy of prototypes: prototypes as filters, prototypes as manifestations of design ideas. ACM Trans Comput Hum Interact 15(2):1–27

Lyon JB (2011) Balancing the emic and the etic: an ethnographer of design reflects on design ethnography. Innovation 30:26–30

Ossher H, van der Hoek A, Storey M-A, Grundy J, Bellamy R (2010) Flexible modeling tools (flexitools2010). In: Proceedings of the 32nd ACM/IEEE international conference on software engineering, ICSE'10, vol 2. ACM, New York, pp 441–442

Özcan M, Parry P, Morrey I, Siddiqi J (1998) Visualisation of executable formal specifications for user validation. In: Margaria T, Steffen B, Rückert R, Posegga J (eds) Services and visualization – towards user-friendly design, vol 1385, Lecture notes in computer science. Springer, Berlin/Heidelberg, pp 142–157

Pohl K (1993) The three dimensions of requirements engineering. In: CAiSE'93: Proceedings of advanced information systems engineering. Springer, London, pp 275–292

Pohl K (2010) Requirements engineering: fundamentals, principles, and techniques. Springer, Heidelberg/New York

Rensink S (2010) The edge of graph transformation graphs for behavioural specification. In: Engels G, Lewerentz C, Schäfer W, Schürr A, Westfechtel B (eds) Graph transformations and model-driven engineering, vol 5765, Lecture notes in computer science. Springer, Berlin/Heidelberg, pp 6–32

Richter S (2011) Gesteuerte und interaktive Simulation von virtuellen Prototypen. Master's thesis, Hasso Plattner Institute at the University of Potsdam, Germany

Snyder C (2003) Paper prototyping: the fast and easy way to design and refine user interfaces. Morgan Kaufmann, San Francisco

Stachowiak H (1973) Allgemeine Modelltheorie (German). Springer, Wien

Sutherland M, Maiden N (2010) Storyboarding requirements. IEEE Softw 27:9–11

Teusner R (2011) Smarte Simulation von virtuellen Prototypen. Master's thesis, Hasso Plattner Institute at the University of Potsdam

Tietz D (2011) Sichtbasierte Transformationen von Anforderungen in verschiedene Repräsentationen. Bachelor's thesis, Hasso Plattner Institute, University of Potsdam

Tohidi M, Buxton W, Baecker R, Sellen A (2006) Getting the right design and the design right: testing many is better than one. In: CHI'06: Proceedings of the SIGCHI conference on human factors in computing systems. ACM, New York, pp 1243–1252

van der Aalst WMP (2007) Trends in business process analysis: from validation to process mining. In: International conference on enterprise information systems, ICEIS, Funchal/Madeira, 12–16 June 2007

Winograd T (ed) (1996) Bringing design to software. ACM, New York

What Can Design Thinking Learn from Behavior Group Therapy?

Julia von Thienen, Christine Noweski, Christoph Meinel, Sabine Lang, Claudia Nicolai, and Andreas Bartz

Abstract Some widely-used approaches in Behavior Group Therapy bear a striking resemblance to Design Thinking. They invoke almost identical process-models and share central maxims like "defer judgement" or "go for quantity". Heuristics for composing groups (mixed!) and preferred group sizes (4–6) are very much alike as well. Also, the roles ascribed to therapists are quite similar to that of Design Thinking coaches. Given these obvious analogies, it is most natural to ask what the two traditions can learn from one another – and why it is that they are so strikingly alike. This article ultimately hopes to inspire further investigations by giving examples of how Design Thinking may profit from taking a look at Behavior Group Therapy. We will discuss (a) new techniques for coaches to detect and treat personal dissonances that impede project work, (b) new methods for teams to upgrade empathy, find crucial needs or test prototypes and (c) theoretical insights regarding what happens in the process.

The fact that Design Thinking and some widely-used approaches in Behavior Group Therapy are so strikingly similar that one could almost pass for the other may seem peculiar. In any case, it is a wonderful opportunity to learn from one another.

Part I of this article gives a short synopsis of what is actually so similar in Behavior Group Therapy, from the point of view of Design Thinking. *Part II* introduces several techniques widely used in Behavior Group Therapy which may be of value for design thinkers. *Part III* borrows analytic means common in Behavior Therapy to nourish Design Thinking theory.

J. von Thienen • C. Noweski • C. Meinel (✉) • S. Lang • C. Nicolai • A. Bartz
Hasso-Plattner-Institut (HPI), für Softwaresystemtechnik GmbH,
Prof.-Dr.-Helmert-Street 2-3, Potsdam 14482, Germany
e-mail: meinel@hpi.uni-potsdam.de

H. Plattner et al. (eds.), *Design Thinking Research*, Understanding Innovation,
DOI 10.1007/978-3-642-31991-4_16, © Springer-Verlag Berlin Heidelberg 2012

Due to limited pages and the overarching subject of this volume, our investigation unfortunately will remain one-sided. The complementary question of what Behavior Group Therapy can learn from Design Thinking is certainly just as auspicious, and we hope to pursue it elsewhere. Suffice it to say here that, as a general impression, Design Thinking appears more straightforward and polished than its sibling procedure in Group Therapy. Design Thinking could almost pass off as the advertisement or movie version of an otherwise common story. Its core ideas have been prepared to reach an unmatched degree of clarity. Every heuristic is carried out with a unique determination to be consistent up to final details, the pace is dizzying and the whole enterprise tends to come across as strikingly entertaining. These qualities surely are desirable in other domains too, such as in Behavior Group Therapy.

1 Part I: What Design Thinking and Basic Approaches in Behavior Group Therapy Have in Common

When saying that "basic approaches" in Behavior Group Therapy bear a striking resemblance to Design Thinking, the likely question to follow is: Which therapeutic approaches are we talking about?

In Behavior Therapy, there are generally two types of groups. In one case, the group runs through a pre-specified program, e.g., a program to improve social skills, self-control or stress management. In the second case, there is no pre-specified program and people with varying problems may attend. How these latter groups work is what we will be talking about.

1.1 The Process

All in all, Behavior Therapy is highly problem-oriented. The process of identifying a problem that client and therapist agree to work on and then trying to solve it is omnipresent in Behavior Therapy. Of course, just like design thinkers structure their problem solving process, behavior therapists too have more on offer than a fishing expedition into the blue. "'Typically behavior therapy' was and is, that behavior therapists work according to the pattern of a structured problem-solving process" (Fiedler 1996, p. 48, o.t.[1]).

This process comprises a number of distinct phases that build on one another. To solve a problem, you go through one phase after the other. Or you return to an earlier phase and iterate if results don't seem satisfactory yet.

---

[1] Here and in what follows the abbreviation o.t. means "our translation".

Design thinkers know how time can bring a multiplicity of process formulations, differing in the number of phases differentiated and in the names given to them. Yet, what the diverse models ask you to do remains basically the same. Why should things be all that different in Behavior Therapy?

> Even though several models exist which invoke a varying number of phases in the problem solving process, their content is typically very similar. Over the years, a unified model has emerged. Since the early 1980s, proceeding according to a process with six phases is regarded as necessary and sufficient for therapeutic problem solving. (Fiedler 1996, p. 48, o.t.)

Figure 1 gives an overview, comparing the model of problem solving used in Behavior Group Therapy to a common model of Design Thinking.

In Behavior Group Therapy, this is what you do in the six phases of problem solving. . .

1. **Phase I:** In the beginning, you collect as much information as you can regarding the domain where something is at odds somewhere. The exploration should yield a panoramic overview and it should allow you to take on different perspectives when looking at the domain of interest. Eventually, the aim of phase I is to develop a sense of what the problem actually is, at its core.
   In Behavior Group Therapy, one client (the "focus client") contributes a personal problem that the whole group will work on afterwards. Typically, the problem domain is explored by interviewing the focus client; the other group members do the interviewing. Another common tool is role-playing: Problematic situations are staged in the therapy setting. Sometimes the problem is "materialized" by setting up a sculpture. For example, the client arranges his group mates to represent important others (spouse, boss, colleagues. . .) in whatever respects are important (emotions, demands they make, conflicts. . .).
2. **Phase II:** Now that the problem has been understood, it is time to put it in a nutshell: Name it! Since there are typically alternative options, choose a focus that is worth pursuing.
3. **Phase III:** The next step is to generate multiple options for solving the problem. Typically, this is done via brainstorming. Of course, this means going for quantity and deferring judgment. Osborn's (1953) classical brainstorming rules are alive and well.
   In Behavior Group Therapy, it is common to have the rest of the group brainstorm while the focus client is asked to debark in this stage of process (Bartz 2011), thus yielding something like a split between "design team" and "user".
4. **Phase IV:** One possible approach is picked out to be pursued further.
   In contrast to Design Thinking, it is often the focus client (the user), not the brainstorming team who picks an option.
5. **Phase V:** The chosen approach is tried out.
   In Behavior Group Therapy, this phase generally includes staging a role-play: The focus client tries out how things evolve with the approach chosen for testing. Or someone else takes on his role so that he can watch. If time allows and the

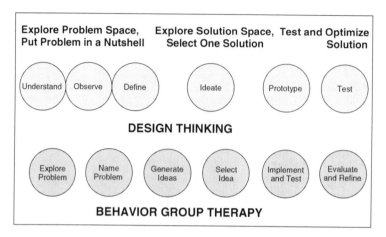

Fig. 1 Comparing the problem solving processes of Design Thinking and Behavior Group Therapy

solution seems promising, the focus client will test his solution in real life as well.

Sometimes it is not easy for the focus client to realize an approach immediately. Then, it is usual to schedule a number of tests with increasing levels of difficulty. This technique is called "hierarchization". Occasionally, it has the positive side-effect that the focus client develops more self-confidence and loosens up considerably. He may even become so courageous as to decide on testing "a wild idea" which is generally considered promising by everyone, but which the focus client himself did not dare to try earlier on.

6. **Phase VI:** Does the approach make things better? Is the solution *subjectively satisfying* for the focus client? Phase VI schedules an evaluation of test results. If insufficiencies become apparent, you go through the problem solving cycle once again, or partially so: Return to whatever phase seems suitable.

Obviously, the process used in Behavior Group Therapy is highly similar to the one used in Design Thinking. Historically, this is probably the case because both traditions allowed themselves to be inspired by the same models or theories of problem solving which became so popular in the early 1960s. In psychology, they received further fine-tuning by the Psychology of Though Processes in the 1960s and 1970s and were adapted early on by Behavior Group Therapy (see Fiedler 1996, for a historic overview).

But **WHY** should Design Thinking and Behavior Group Therapy both have an interest in using this kind of a process? The coincidence may seem less surprising, in fact even logical, when considering the common aim that is at stake. In both cases we do not only want to solve problems, but problems of a particular kind – which have come to be called "wicked" in technical literature (Rittel and Webber 1973). There are several common characteristics of these problems, which are readily identifiable in most problems of Design Thinking or Behavior Group Therapy.

Typically, in the beginning it is unclear what the crux of the matter actually is. Moreover, there is no single correct statement of what the problem is. Rather, one may consider quite different outlooks on the problem. Alternative problem statements are not true or false, they just seem to be more or less fortunate starting points for creating solutions.

For instance, in Behavior Group Therapy a client may complain about her husband who is mostly away from home due to his workload. When he is at home, he is too tired to participate in family life and refuses to consider his wife's arguments, which she articulates whenever he is available for her. What's the problem? Does he need a family life that feels more like being on holidays than like working at a grievance hotline? That is, should family life become so attractive and easy to access that even the overworked husband will happily participate? And/ or does the client need to feel good about her life regardless of what her spouse does? Perhaps she needs an effective way to change her husband ... or to find a better one? Obviously, there is ample scope for framing the problem. And another central characteristic of wicked problems comes into play: The way you decide to phrase your problem predetermines to a large extent what kinds of solutions you will end up with, e.g., means to involve or get rid of the husband.

1.2 Team Size and Composition

Here is another issue that makes Design Thinking and Behavior Group Therapy something like sibling endeavors: In both cases you set up groups or "teams" who will explore a problem and develop solutions. So you will have to decide how many people go in one group and how to mix the teams. Thus, it is not surprising that in both traditions there has been quite a bit of research on issues regarding sensible group setups. Fiedler (1996) gives an overview of research results from Behavior Group Therapy, Bartz (2011) includes comments on the current practice.

Just as in Design Thinking, in Behavior Group Therapy small teams of about four to six people are considered favorable; ten would be an absolute maximum. In addition, it is recommended to adjust the number of group members to the expected intensity of the exchange and the privateness of issues: The more intense and private the expected exchange, the smaller the groups should be.

When it comes to selecting members for a group, the go-for-heterogeneity-rule is a common maxim in Behavior Group Therapy, just as it is in Design Thinking. Notedly, diversity is aspired regarding (a) age, (b) gender, (c) profession and (d) educational background – since these typically go along with certain perspectives which all should inform the group's panoramic view on problems (Bartz 2011).

But there are two noted exceptions to the heterogeneity-rule as well: interest and mental or verbal ability (Bartz 2011). Everyone should be interested in pursuing the group work with the given challenges that there are. For example, when a client has already endured ten meetings of a mental illness group but he himself wants to work on the subject of drug addiction better put him in an addiction group than in another mental illness one if there is a choice. Also: The mental or verbal abilities of participants should make it possible to discuss the subjects that come up in the group. Thus, it is difficult to integrate people who have a hard time even understanding everyday conversations.

But these are all just rules of thumb. In the light of our present-day knowledge, the literature seems to indicate that investing time and effort into balancing-out teams meticulously is rather unwarranted (Bartz 2011).

1.3   The Role of a Therapist or Coach

Design Thinking and Behavior Group Therapy use the same kind of process to gain a fresh outlook on problems and arrive at gratifying solutions. They use similar heuristics for compiling groups. And there is a third commonality which calls for lively exchange: In both traditions the groups enjoy the support of a facilitator. In Design Thinking, that is a teacher or coach. In Behavior Group Therapy it is the therapist.

In all cases, the facilitator appears as an expert of method, not as an expert of ready solutions. He introduces the process model, provides structure and keeps time – but all along he tries hard to avoid doing the team's work for them. The team needs to understand that a certain problem view with its corresponding solution space is actually *theirs*.

One responsibility that is unequally present in the two traditions concerns group dynamics. In Behavior Group Therapy, it is an important job of the facilitator to monitor group dynamics constantly and to come up with stabilizing interventions if needed. While this difference in liability is hardly surprising given the respective clientele, it leaves ample room for design thinkers to check if certain therapeutic techniques might be useful tools for coaches or teachers as well.

2   Part II: What Techniques Design Thinking May Pick Up from Therapeutic Settings

Having explored some of their outstanding commonalities, how can Design Thinking profit from looking at Behavior Group Therapy? We will quickly go through a few techniques which may enrich the Design Thinking method case, starting with techniques for facilitators and then proceeding to techniques for teams.

2.1   Interventions for Facilitators to Stabilize Teams

Having said that, it is considered an important responsibility of the therapist to monitor group dynamics and intervene supportively, it is a matter of course that behavior therapists know reams of stabilizing interventions.

WHY might it be a good idea to have the facilitators stabilize teams in Design Thinking? Oftentimes, design thinkers may be able to stabilize each other in the case of disturbances. If this is the case – great! But if teams don't stabilize themselves, interventions by the facilitators may be crucial: Every team member should be able to save his skin in Design Thinking. Also, it may be easier for external observers to recognize disturbances than for the involved team members themselves. And interventions may be less critical when they come from a neutral person.

Here is a small selection of stabilizing interventions taken from Behavior Group Therapy:

HOW to detect disturbances early on?

There is a strategy that may seem trivial, but it is highly effective: Monitor emotions! All the time! Once negative emotions towards other team members appear, a disturbance is building up.

HOW to intervene in case of dysfunctional group dynamics?

The autonomy of a team needs to be respected. If you feel an intervention is necessary: Ask for permission! For instance: *May I share a personal impression?* ... Await positive reply ... *It seems to me there is something bothering the team that has little to do with the challenge* ... Then, it may help to elucidate needs. People often request or argue against certain concrete measures which they feel are crucial to personal needs. But there may be alternative measures to secure the needs at stake. Teams can proceed more purposefully once crucial needs have been communicated. For example, there may be many reasons why a team member rebels against taking on a certain job. He may feel passed over and wish for more control of his own agenda. He may feel pessimistic about the outcome, needing more certainty that his efforts will have some pay-off. Or he may simply be afraid of visiting certain places or people. As always, solutions vary with the need that is at stake. Try throwing a suggestion into the ring ... *Maybe there is something you need that should be allowed for(?)*.

HOW to handle problematic demands?

Sometimes team members make requests that overstep legitimate boundaries of others. For instance, someone may invoke his need for being in charge of his own agenda, thus trying to claim for himself a decision-making authority which trumps that of his team mates. In this case, a careful intervention is to split your personal view. On the one hand: Validate the need that is at stake – make clear that you consider it an important and justified need. On the other hand: Name the problematic side of the explicated demand. (For example, as follows: *On the one hand, I can certainly understand that you want to be in charge of your own agenda. On the other hand, it seems to me the request we are talking about interferes with the autonomy of others. Maybe there is a procedure that unequivocally grants the same rights to all team members...*)

HOW to bolster team members in case of major offences?

When emotions boil high, sometimes things are said that seriously threaten someone's well-being. In this case, there is a small kind of intervention that often has a major stabilizing effect. You can turn to the offended person and ask a question of the following kind: *Can you bear that ... [name] ... just articulated a view which disagrees with yours?* Thus, you indirectly present a stop sign to the rest of the group saying: Careful, what is going on is hardly bearable for one of you. Secondly, by asking the offended team member if he can bear the situation you take his agony seriously. That typically has a great relieving effect on its own. Finally, a neutralizing description of what happens makes it easier for the team to find a productive and calm way of continuing the discourse. Just imagine you would have decided on a pejorative formulation of the original question: *... can you bear that ... [name] ... just showed his lack of social competences once again and lost himself in complete nonsense?*

HOW to handle "weak" team members?

Sometimes, there is a "weak" team member who might drift off into the role of an outsider, gadfly or scapegoat. Ample research has been carried out about how such dynamics typically affect groups. And the implications are clear: It is a cardinal mistake to let someone become an outsider, gadfly or scapegoat! If that happens, it is not only tragic for the person concerned but for the whole group. Everyone learns: This is not a safe place! Be cautious or otherwise you too may fall through the cracks! Thus, people will censor themselves much more rigorously then they otherwise would. They will tend to avoid self-disclosures, refrain from taking risks or even try to secure themselves by cliquism. If you are a facilitator, your maxim needs to be: Always support weak team members! Avail yourself of all the techniques mentioned thus far! And, in particular, help weak team members in case of major offences!

Most of the mentioned techniques have been adopted from Roedinger (2011) who is extremely rich in these kind of suggestions – thus a good source for everyone whose interest has been sparked.

2.2   Plan Analysis: A Technique to Carve Out Basic Needs

Having considered techniques for facilitators, what is there to pick up for teams?

A tool that may be interesting for many reasons is *plan analysis* which is something like an elaborate version of why-how laddering. Generally, it is a technique to upgrade empathy.

WHY use plan analysis? Because it helps ...

- To find out why a person behaves and feels the way he or she does,
- What her ultimate needs are,

- Where there are conflicts between needs and reality or between several aims a person has and
- What alternative behaviors (which the person does not show yet) would be viable for her.
- Last but not least, it is a technique that sharpens an investigator's empathy to a degree where it becomes almost analytic.

To understand how plan analysis works, a concrete example will be helpful. So let's return to the client from Behavior Group Therapy who complains about her neglectful husband: He is hardly ever at home. He does not participate in what remains of their family life. But the charges uttered by our client trail off unheard, or so it seems.

HOW to perform a plan analysis? You spell out a hierarchy of strategies: At the lowest level, you name concrete behaviors, formulated in the indicative. Above that, you place ever more general maxims, strategies or plans, phrased in the imperative. Ultimately, such plans point to basic needs that are at stake for the person.

The woman who complains about her husband may follow a superordinate plan saying: "Avoid being alone!", because she feels helpless when she is on her own. To sidestep this dreaded situation, she may have acquired a subordinate plan like: "Create obligations!". This she may try to achieve by following the plans "Be married!" and "Claim support!". In consequence, when her husband is present she typically enumerates his obligations and blusters when he fails to provide the expected support (see Fig. 2).

Unfortunately for the client, the effects of her behavior are adverse to her superordinate plan "Avoid being alone/helpless!". Rather, by enumerating obligations and blustering she drives her husband further and further away.

The conflict between aims or plans and real-life-behavior-consequences is indicated by grey flash lines in Fig. 2. (Using such flash lines is not common in plan analysis so far, but we feel it is a useful amendment.)

As always, what is one person's poison may be another one's meat. While the client is unhappy, both therapist and design thinker have identified needs that could provide worthy starting points for problem-solving: There are unsatisfied needs of feeling integrated, safe and in control.

It is worth noting that concrete behaviors alone hardly reveal what needs and plans are at stake. Thus, the very same client, blustering and reproachful as she is, could very well follow quite different plans to those just spelled out for her. And if her plan-need couples were different, surely a design challenge would have to take another turn too. Imagine: It could well be that the client basically tries to get somewhere in life. Everyone including herself is supposed to see how she made it. Her life is supposed to be like a fairy tale come true. Thus, she feels she needs a rich husband, a big house, plenty of kids and, of course, happy family get-togethers. What a different design challenge that would make! Thus, Fig. 3 shows quite a different plan structure. Maybe this is what makes our blustering, unhappy woman tick.

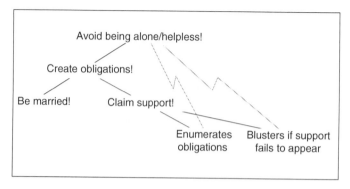

Fig. 2 Plan analysis for blustering woman: "Avoid being alone/helpless!"

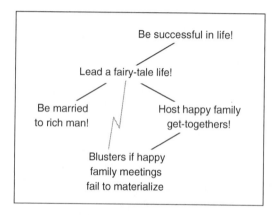

Fig. 3 Plan analysis for blustering woman: "Be successful in life!"

Obviously, observing behaviors alone does not suffice to work out plan structures. Here comes the point where design thinkers may upgrade their empathy. Plan analysis comes with tools which help you broaden and focus your observations when trying to understand someone. These tools are questions which you bare in mind when dealing with a person.

HOW to figure out what plans people follow, what needs they have? Answer the following five questions!

1. **Which emotions and impressions does he/she elicit in me?**
   In the case of our blustering woman, imagine how she would come across differently depending on whether (a) she wants to create obligations to avoid the helplessness of being alone or (b) she wants to come across as a fairy tale queen.
   When dealing with the support-yearning women, you probably feel absorbed, pressurized or overstrained and this may create anger. The fairy tale queen,

however, may appear outstandingly dressy or she might act as a particularly loving mother with stunningly well-turned-out beautiful children. Depending on your own preferences you might marvel or pity her.

2. **How does he/she want me to behave?**
   The helpless woman probably wants you to be supportive and reliable. The fairy tale queen wants to sell her success, she may want you to admire, celebrate or envy her.

3. **Which behavior tendencies does he/she actually elicit in me? How am I inclined to behave towards him/her?**
   In case of the support-yearning woman you might want to withdraw in the face of her effervescent demands. In case of the fairy tale queen you might wish to have a serious word with her.

4. **What image of himself/herself does he/she try to convey?**
   In the first case, the woman wants to come across as being entitled to raise claims. In the second case, well, she wants to come across as successful, as a fairy tale queen.

5. **Which behavior on my part is he/she trying to avoid?**
   If the woman wants help, she probably would not want you to question her right to claim support. And she would try to avoid that you leave her alone. If the woman wants to feel successful, she would try to avoid that you think she is just anybody, insignificant, a loser, someone you can forget about.

Thus, depending on the plans people follow and the needs they have the five questions will be answered differently – even if immediate behaviors seems very similar. So the questions help to carve out plans and arrive at crucial needs.

If you want to learn more about plan analysis, take a look at Caspar (2007).

2.3 Chair Dialogue: A Technique for Testing Prototypes

A second technique that may be of interest for design thinkers is the *chair dialogue* to test prototypes.

WHY use the chair dialogue? Because sometimes a user is uncertain, ambivalent or hesitant regarding a prototype – but he has a hard time explicating his impressions. In this case, the chair dialogue may help him to deliver clear and capacious thoughts.

HOW to carry out a chair dialogue? Arrange two empty chairs for the user. One chair, you explain, belongs to the sceptic or opposer of the prototype. The other chair belongs to the supporter of the tested approach. Then, you let the person say what comes to his mind as he sits on one chair after the other and takes on the corresponding roles. Additionally, the members of the design team can engage both opposer and supporter of the idea into a conversation or interview.

Of course, members of the design team can also sit on the two chairs and empathically take on the roles of a user doubting or loving the solution. This may help you sharpen your understanding of what is valuable, repelling or missing in a prototype.

3 Part III: How Therapeutic Theory May Inspire Design Thinking Theory

Now that the look at Behavior Group Therapy has yielded a couple of ideas that may (or may not) be of interest for facilitators and design thinkers, what does Behavior Group Therapy have to offer for Design Thinking researchers?

Indeed, the prospects for expanding on Design Thinking theory are good: Next to the sweeping parallels in practical procedures, Behavior Therapy brings along an elaborate corpus of theoretical reflections.

Here are some bits and pieces of theory we consider particularly interesting because...

- They provide frugal means to make sense of variegated practices in Design Thinking,
- They may help to invent new valuable practices and
- They may be useful for students to make sense of the "Design Thinking universe" they newly encounter.

3.1 Analyzing Settings: Security as a Basis for Innovation

For many decades, research has shown how there is an immediate link between feeling secure on the one hand and explorative behavior or contributing something creative to the world on the other hand (Bowlby 1988; Holmes 1993; Maslow 1954).

In a therapeutic process, a lot is to be innovated. People generally will acquire a new outlook on old problems. In addition, they will have to try out completely new strategies to ultimately reach new and workable solutions.

At the same time, great uncertainty may prevail – and it can be quite bothersome. Oftentimes, clients don't know what to expect. They have no idea what insights await them regarding the problem. They can't tell whether a viable solution will indeed be found in the end and what it will demand of them personally.

Such a degree of uncertainty is not only scary for clients in therapy; it is scary for basically everyone. So there is a tension between insecurity on the one hand and a need for innovation on the other.

WHY does insecurity impede innovation? When people feel insecure they will tend to stick with the familiar because it provides a sense of safety. Thus, insecurity is a powerful blocker of innovation.

HOW to overcome familiarity-clinging, how to make innovation possible? You provide a setting which is so safe, it is almost artificial. There, people can engage with change. Key factors are (a) social relations which need to be reliable and supportive, (b) powerful means to tame draw-backs or criticism and (c) bolstered convictions regarding oneself: that one can actually handle the challenges no matter what comes.

Surely, the accordance with schools of Design Thinking is obvious. Here too students may be bothered by uncertainties. But fortunately for them and the innovations to come, their learning environment is an enormously safe setting in many regards. But let's consider one issue at a time, comparing the therapeutic setting to schools of Design Thinking.

Social Support. With respect to social relations, in Behavior Group Therapy it is the job of the therapist to reliably support every group member, particularly when a person is needy, and to ensure that conversations in the room take a constructive turn. In Grawe's (2004) words, the process is best supported if the therapist comes across as "sensitive-empathic, understanding and accepting, engaged in the well-being of the client, as trustworthy and dependable, as warm and supportive and as competent" (p. 404, o.t.). In addition, the therapy setting induces dependability as the meetings will takes place predictably within a known period of time.

Clearly, there are analogies to schools of Design Thinking which go far beyond predictable schedules. Here too the facilitators establish a culture of conduct which is characterized by mutual benevolence, respect, curiosity, approval and support. For example, there are rituals of clapping warm-ups to boost exchange and to help build on each other's ideas, there are non-competitive presentations with constructive feedback sessions etc. Obviously, schools of Design Thinking are set up in such a way as to be socially very safe places.

Failure. Crucial factors that may all too easily destabilize a person's sense of security and wellness are "criticism" and "draw-backs". To secure readiness for change, experiences of that kind need to be tamed and they need to become bearable. Both in Design Thinking and in Behavior Group Therapy it is a likely manoeuvre to shed light on their constructive side. In Design Thinking, there are multiple strategies for doing this. Mottos like "fail early and often" emphasize the positive force of personally challenging experiences. Common formats such as "I-wish-I-like" instead of "I-dislike" automatically cast feedback into constructive forms. Notably, negative feedback is not being downplayed or discredited this way. Quite the contrary, it is being embraced – by asking what value it contains. Thus, negative experiences are being tamed. They can be handled and learned from. People just need the strength to confront them without impermeable shields of self-defense.

Self-efficacy. Finally, both people in Behavior Group Therapy and in Design Thinking need a healthy sense of self-efficacy to fully engage in their challenges. If they doubted success was possible for them, why should they engage? But even for the halting minds self-efficacy is easily gained either in Behavior Group Therapy or in Design Thinking. First, there is the rock-solid confidence on the part of the facilitators who obviously believe that something worthy can be reached. Then, there are good experiences with the process: Even if you don't know where the process will lead you while you are on the way, time after time things seem to turn out well in the end. And if they don't, you can always go back and iterate the process until you are satisfied.

3.2   Mindset Analysis

Now that settings have been analyzed in an important regard – they need to be socially safe places to allow for innovation – let's approach the people who populate it and try to take a trip right into their minds. Behavior Therapy comes with a tool to *analyze mindsets* that may be quite valuable for Design Thinking research. And it may help to see how these mindsets, just like the interpersonal settings, convey a notion of safety needed for innovation.

WHY analyze mindsets? Because mindsets govern variegated behaviors and emotions. Thus you can explain, predict and generate a lot (of behaviors or emotions) with a little (mindset analysis).

HOW to analyze mindsets? You spell out central cognitions in the form of belief-sentences.

A mindset which is pretty much the opposite of what one would attribute to a design thinker was described by Beck (1976). Regarding behaviors and emotion, it is associated with a lack of engagement and a lack of joy: the mindset of a depressed person. It is characterized by negative convictions regarding. . .

1. The self (e.g., "I am incompetent.")
2. The world or environment (e.g., "The word is a hostile place.") and
3. The future (e.g., "Nothing will ever change for the better.").

Obviously, it makes little sense to engage if one is incompetent anyway and if nothing will ever change for the better, regardless of what one does.

In contrast, the mindset of a design thinker needs to bias towards action. Accordingly, belief-sentences will express a high degree of self-efficacy. Central cognitions could be:

1. **Oneself**
   "I can make a change for the better."
2. **The Future**
   "The future is up to me (us)."
3. **The World**
   "The world is like an extended living room."

This last formulation picks up one of our earlier research findings: "Being in one's living room with friends" is a situation that feels much like "being at a d. school" (von Thienen et al. 2012). The latter is a place where you learn and practice Design Thinking. What the two situations have in common is...

- That you are in a relatively safe place,
- Not alone, but with others
- You will have to take their needs into account,
- The others will generally be friendly and inclined to co-operate;
- You can have a good time together.
- Another important issue is that your surroundings are configurable: You can arrange the place according to your wishes and taste, as long as the needs of others are respected as well.
- And as long as your resources last (there are constraints).
- Adjusting your surroundings in a way that makes sense is not some far-fetched possibility but something to be taken for granted;
- You are responsible for how your living room looks and what happens in it together with others.

Thus, "the world is like an extended living room" seems to be a belief-statement which captures many crucial aspects of a Design Thinking world view.

Clearly, "others" are an essential part of the Design Thinking world. Important convictions regarding fellow human beings which may likely be strengthened by Design Thinking education could be...

4. People
"People are a source of cooperation on equal grounds."
"People are understandable. There are reasons for how they act."

Then, there need to be convictions which help to confront the new and handle obstacles. These may be convictions such as the following:

5. Otherness
"Otherness is promising." ...instead of the rather prevalent conviction... "Otherness is threatening."
6. Obstacles
"Problems are welcome occasions."
"Failure is productive feedback."
7. Uncertainty
"Oftentimes, uncertainty paves the way of beautiful options."

In addition, Design Thinking seems to generate convictions which ease life by being anti-perfectionist, anti-rigid and learning-oriented. They reduce the inhibition threshold to become active in the first place, thus strengthening the bias towards action even more. These could be convictions such as:

8. Endings

"All we ever do is prototyping."

Thus, there hardly is an ultimate end. If some prototype turns out to be lacking, you just go over it again. So, you don't have to avoid getting started in the first place, considering how likely you may fail in face of your own perfectionist expectations.

There may be another important conviction which helps you get going:

9. Waste

"Waste is important."

"It is a good idea to head for overspill and selection, overspill and selection."

If your first trial had to be perfect, you would probably waver and deliberate and hesitate. If waste is okay, then what is stopping you? Get going!

Last but not least, there seem to be rather unique convictions regarding games, joy and fun which are quite typical of Design Thinking.

10. Fun

"Fun is an important engine of success." ... instead of an otherwise common belief: "Fun is when you are away from work."

Whether indeed these are crucial cognitions for design thinkers, further research and collegial exchange have yet to show. But if so, one could try to be even more comprehensive in the explication of mottos (such as "fail early and often") or in other means to make the Design Thinking mindset "tangible" and easy to adopt for those who enter a d.school.

3.3 How to Survive with a Design-Thinking-Mindset?

After featuring some therapeutic techniques and theory bites, to ultimately inspire further research in this regard and help widen the Design Thinking universe let us finish with a question. We ask it with respect to Design Thinking, but an analogue consideration also could pertain to Behavior Group Therapy. It could well be a domain of common pondering or best practice exchange.

The question, which may seem provocative, is: How are students to survive with a design-thinking-mindset?

WHY is survival a challenge for design thinkers? Just think about it! The mindset people seem to acquire in Design Thinking is pretty optimistic. ("I can make a change for the better.") A hostile world might all too easily frustrate someone with this kind of optimism and verve. Imagine a design thinker at a place where people

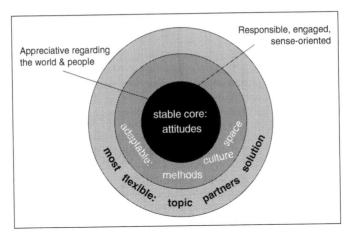

Fig. 4 Cognitive-behavioral model of a design thinker: stable, adaptable and most flexible elements of a well endowment

are supposed to obey orders without expressing their own opinion – a highly traditional company for instance.

What a design thinker is supposed to accomplish seems almost super-human. On the one hand, he is supposed to be optimistic enough to engage with verve in new and unpredictable projects. On the other hand, he is supposed to leave the d.school some time or other. Then he has to leave his almost artificially safe "homeland" where confidence is stabilized by variegated measures. What if the design thinker takes on a problem that happens to be situated in an environment which does not welcome much interference or change?

Obviously, students who learn Design Thinking need to be equipped with some insulation which protects their acquired optimistic outlook from being frustrated or even overridden by "hostile" environments.

We think schools of Design Thinking are already offering much in that regard – and it is illuminating to see it this way, as shown in Figure 4. In between a mindset (which is supposed to be stable) and the concrete action of a design thinker (which is supposed to be highly flexible) comes a protective belt (which is neither completely fixed nor highly flexible; it is adaptable). So here is our suggestion:

HOW to protect the verve and optimism of a design thinker? Provide him with a protective belt that includes methods, culture and habits of using space.

Methods. If you learn and practice how to approach people (e.g., conduct interviews in an inviting manner), chances are people will co-operate and not rebuff you time and again.

Culture. The Design Thinking community entertains a rich culture with many rituals (such as clapping), formats (I-wish-I-like etc.), model arrangements (work with music, games, food. . .), codes of conduct (such as "defer judgment") and a lot more that may be adopted and reproduced. Thus, when students encounter a "hostile" environment they may apply a two-step-strategy. First: Phase in a culture that is Design Thinking friendly! Then: Practice Design Thinking!

Space. In Design Thinking, you are encouraged emphatically to use and adapt the space that there is. While this certainly has many positive effects (e.g., rigidities of thinking may be transgressed, the focus on needs and the bias towards action is promoted anew) it may also be a strategy to survive as a design thinker. Once again, all too often the big, wide world may require a two step process. First: Create the space you need! Then: Live Design Thinking!

With these considerations we hope to have piqued some curiosity regarding what else there is for design thinkers and behavior group therapists to learn from one another. Obviously, we share many central concerns such as solving wicked problems in heterogeneous small groups of 4–6 members by following a common process model. Behavior Group Therapy and Design Thinking have developed a bulk of similar techniques including warm-ups, role-playing, why-how laddering or plan analysis and many more. We share interests in understanding how the process works and what it demands, such as a thorough sense of safety and a mindset that biases towards action, which is supposed to survive even in "hostile" environments. All these common concerns, techniques and research questions – vociferously – call for further exchange.

References

Bartz A (2011) Einführung in die verhaltenstherapeutische Gruppenpsychotherapie. BFA, Berlin

Beck AT (1976) Cognitive therapy and emotional disorder. International University Press, New York

Bowlby J (1988) A secure base: parent–child attachment and healthy human development. Routledge, London

Caspar F (2007) Beziehungen und Probleme verstehen. Eine Einführung in die psychotherapeutische Plananalyse. Hans Huber, Bern

Fiedler P (1996) Verhaltenstherapie in und mit Gruppen. Psychologische Psychotherapie in der Praxis. Beltz, Weinheim

Grawe K (2004) Neuropsychotherapie. Hogrefe, Göttingen

Holmes J (1993) John Bowlby and attachment theory. Makers of modern psychotherapy. Routledge, London

Maslow AH (1954) Motivation and personality. Harper and Row, New York

Osborn A (1953) Applied imagination: principles and procedures of creative problem solving. Charles Scribner's Sons, New York

Rittel H, Webber M (1973) Dilemmas in a general theory of planning. In: Policy sciences, vol 4. Elsevier Scientific Publishing Company, Amsterdam, pp 155–169

Roedinger E (2011) Praxis der Schematherapie. Lehrbuch zu Grundlagen, Modell und Anwendung. Schattauer, Stuttgart

Thienen Jv, Noweski C, Rauth I, Meinel C, Lang S (2012) If you want to know who you are, tell me where you are: The importance of places. In C Meinel & L Leifer (Eds.), Understanding innovation. Design thinking research. Studying co-creation in practice. Springer, Berlin